4th International Conference on X-Ray and Electrovacuum Techniques 2017

Journal of Physics: Conference Series Volume 967

Saint-Petersburg, Russia
23 – 24 November 2017

ISBN: 978-1-5108-5853-4
ISSN: 1742-6588

Printed from e-media with permission by:

Curran Associates, Inc.
57 Morehouse Lane
Red Hook, NY 12571

Some format issues inherent in the e-media version may also appear in this print version.

Printed by Curran Associates, Inc. (2018)

For permission requests, please contact the Institute of Physics at the address below.

Institute of Physics
Dirac House, Temple Back
Bristol BS1 6BE UK

Phone: 44 1 17 929 7481
Fax: 44 1 17 920 0979

techtracking@iop.org

Additional copies of this publication are available from:

Curran Associates, Inc.
57 Morehouse Lane
Red Hook, NY 12571 USA
Phone: 845-758-0400
Fax: 845-758-2633
Email: curran@proceedings.com
Web: www.proceedings.com

Table of contents

Volume 967

4th International Conference on X-ray and Electrovacuum Technique

23–24 November 2017, Saint-Petersburg, Russian Federation

Accepted papers received: 1 February 2018
Published online: 15 February 2018

Preface

Papers

IOP Conf. Series: Journal of Physics: Conf. Series **967** (2018) 011001 doi:10.1088/1742-6596/967/1/011001

4th International Conference on X-ray and Electrovacuum Technique – New Ideas and Developments

N N Potrakhov, A Yu Gryaznov, A A Lisenkov and D K Kostrin

Department of electronic instruments and devices, Saint Petersburg Electrotechnical University "LETI", 197376, Saint Petersburg, Russia

E-mail: kzhamova@gmail.com

Abstract. In this preface key features of the 4th International Conference on X-ray and Electrovacuum Technique that was held on 23–24 November 2017 in Saint Petersburg, Russia are discussed.

On 23–24 November 2017 the representatives of the leading manufacturers of X-ray equipment, universities and research organizations gathered in the walls of the Saint Petersburg Electrotechnical University "LETI" to discuss the latest ideas, developments and achievements in the industry.

The organizer of the 4th International Conference on X-ray and Electrovacuum Technique was Saint Petersburg Electrotechnical University "LETI" with the support of CJSC "Svetlana-X-Ray", OJSC SPC "Burevestnik", CJSC "ELTECH-Med", Saint Petersburg Scientific and Research Institute of Radiation Hygiene n. a. Professor P. V. Ramzaev, Moscow State Technical University of Radioengineering, Electronics and Automation, All-Russian Scientific, Research and Test Institute of Medical Equipment, CJSC "SRPC Elektron", CJSC "AMICO", LLC "Diagnostika-M", CJSC "Medical Technologies Ltd", LLC "Spektroflesh", LLC "Sintez NPF", LLC "Flash electronics".

The conference was attended by 48 experts from the technical universities and industrial enterprises: Saint Petersburg Electrotechnical University "LETI", LLC "Spektroflesh", LLC "CB X-ray devices", CJSC "Svetlana-X-Ray", LLC "Kurs-AS1", LLC "Pulse technologies", LLC "Ekroskhim", LLC "SPC "SPECTRON", CJSC "Medical Technologies Ltd", FSUE "SRI SPC "LUCH", JSC "CSRI "Electron", SBEHC "Scientific-practical center of medical radiology DZM", JSC "Vimatek", CJSC "Sintez NDT", JSC "Scientific instruments".

In his welcoming speech the Vice-Rector for Research of Saint Petersburg Electrotechnical University "LETI" Dmitrii V. Gaivoronskii wished the participants of the 4th International Conference on X-ray and Electrovacuum Technique mutual understanding, interesting solutions to current problems, and new contacts and business partners.

"The issues that will be discussed today cover not only medical subjects, but also flaw detection, data transfer in the X-ray frequency ranges. The Conference has already passed the period of formation, has grown to the scale of International" – Vice-Rector for Research of Saint Petersburg Electrotechnical University "LETI" Dmitrii V. Gaivoronskii.

In his report head of Department of electronic instruments and devices of Saint Petersburg Electrotechnical University "LETI" Nikolay N. Potrakhov told about the visit of the team of specialists from Saint Petersburg Electrotechnical University "LETI" to the World's Fair of Medical Equipment MEDICA 2017, which took place from 13 to 16 November 2017 in Dusseldorf

IOP Conf. Series: Journal of Physics: Conf. Series **967** (2018) 011001 doi:10.1088/1742-6596/967/1/011001

(Germany). The participation of the delegation of the Saint Petersburg Electrotechnical University "LETI" became possible thanks to long-term cooperation of the University with the Committee on Industrial Policy and Innovation of Saint Petersburg. At the exhibition was presented one of the latest developments of Saint Petersburg Electrotechnical University "LETI" performed in cooperation with a small research and development company CJSC "ELTECH-Med" – mobile X-ray complex "PARDUS-Stoma". The device caused an increased interest of foreign experts – the Minister of one of the lands of Germany for more than two hours talked with representatives of Saint Petersburg Electrotechnical University "LETI".

During the Conference (figure 1) were presented more than twenty speeches by heads of leading Russian enterprises specializing in design and manufacture of X-ray equipment. The presentations covered both theoretical issues and examples of practical applications of X-ray and electrovacuum equipment in a range of medical industries – neonatology, brachytherapy and dentistry, as well as in industry and agriculture.

Figure 1. Some photos from the Conference.

"Participants thank us for that we have revived the tradition of holding Conferences on X-ray technology. In Soviet times, there were many events to discuss the latest advances in our industry, but after the collapse of the Soviet Union, such meetings were not held for over 20 years. Next year we plan to invite more foreign participants to our fifth Conference, and also be part of the event on the basis of CJSC "Svetlana-X-Ray" – Head of Department of electronic instruments and devices of Saint Petersburg Electrotechnical University "LETI" Nikolay N. Potrakhov.

The Conference has become an effective communication platform on which were discussed the most promising research works in the development of new X-ray instruments, was generalized experience of introduction and exploitation of new models of domestic X-ray equipment, outlined prospects of development of production of X-ray and electrovacuum equipment in Russia.

Peer review statement

All papers published in this volume of *Journal of Physics: Conference Series* have been peer reviewed through processes administered by the proceedings Editors. Reviews were conducted by expert referees to the professional and scientific standards expected of a proceedings journal published by IOP Publishing.

IOP Conf. Series: Journal of Physics: Conf. Series **967** (2018) 012001 doi:10.1088/1742-6596/967/1/012001

Non-destructive inspection of polymer composite products

A N Anoshkin[1], A F Sal'nikov[1], V M Osokin[1], A A Tretyakov[1], G S Luzin[1], N N Potrakhov[2] and V B Bessonov[2]

[1] Perm national research polytechnic university, 614000, Perm, Russia
[2] Saint Petersburg Electrotechnical University "LETI", St. Petersburg, Russia

E-mail: osokin-kt@pstu.ru

Abstract. The paper considers the main types of defects encountered in products made of polymer composite materials for aviation use. The analysis of existing methods of nondestructive testing is carried out, features of their application are considered taking into account design features, geometrical parameters and internal structure of objects of inspection. The advantages and disadvantages of the considered methods of nondestructive testing used in industrial production are shown.

In connection with the rapid growth of the technological development of the aviation industry, extensive introduction of polymer composite materials (PCM) occurs in both lightly loaded and heavy loaded aircraft products. This is due to the fact that some of the metallic units do not meet the technical requirements for specific characteristics, which are presented by prospective aircraft propulsion systems. In addition, the use of composite materials can significantly reduce weight, improve the environmental parameters and economy of the aircraft. The use of PCM in modern technology, especially in aircraft, requires consideration of such factors as anisotropy of rigidity and strength, the non-linear nature of deformation under loading, the probability of destruction along the interface of the layers, and the possible presence of specific internal defects of various types. Among the many defects that arise in PCM, the most frequent ones can be identified, such as: pore defects, interlayer delaminations, peeling of adjacent packets, jammed layers, resin pockets (figure 1), detachment of shells from cellular structure in a multilayer structure (is used in products of sound-absorbing structures) and others [1].

All this diversity can be divided into two large types: operational and production-technological. Operational defects arise under the influence of cyclic loads, the consequences of the human factor, as well as in connection with natural phenomena, such as lightning and others. Industrial and technological, in turn, are divided into chemical, mechanical and geometric. At the same time, one of the main defects that occur at all stages of the product life cycle are bundles and delaminations. They are of decisive importance when deciding on the use of composites in a product. Along with this, the main places for the appearance of delaminations are stress concentration zones, for example, places with an excess of layers of reinforcing material or a sudden change in thickness. The appearance of these defects most often leads to the destruction of the element, at lower loads than their calculated values.

At present, there is a wide range of methods and devices for nondestructive testing (NDT) used for flaw detection of PCM products. This is due to the variety of design features of control objects, their geometric parameters, internal structure, which affects the possibility of using and reliable detection of defects by NDT methods. The main NDT methods used to control the quality of PCM products are:

IOP Conf. Series: Journal of Physics: Conf. Series **967** (2018) 012001 doi:10.1088/1742-6596/967/1/012001

acoustic, thermal and radiation. These methods, in turn, can be divided into different subgroups, which have their advantages and disadvantages.

Figure 1. Visible defects in the design of PCM flanges: *1* – delamination; *2* – resin pockets; *3* – pores; *4* – folds.

One of the most capacious groups is the acoustic methods of NDT. Considering this group, it is easy to see that of all acoustic methods the low-frequency frequency with frequency not exceeding 1–5 MHz is most often used. Its application is due to strong attenuation and dispersion of high-frequency ultrasonic waves in the matrix, fibers and various inclusions of PCM. The main area of application of the method is the detection of interlayer delamination zones in the layered structure of composite material. At the same time, the main advantage of the method lies in realizing control over practically all materials and structures. The requirement for the curvature of the product is determined solely by the dimensions of the sensor, which ensures the same quality control of concave and convex parts.

Also at the present time the ultrasonic method on the phased array is becoming increasingly popular. It allows you to control complex objects. Unlike traditional ultrasound methods, in which the result is a single A-scan, the phased array method is a set of A-scans. The number of A-scans depends on the number of piezoelements in the converter. Further, the set of A-scans makes it possible to visualize the structure of the material, highlighting the frequency of the amplitude oscillations by converting it into an S-scan (figure 2). At the same time, the minimally detected defect is significantly smaller in comparison with the traditional method. This is due to the smaller size of the piezoelements.

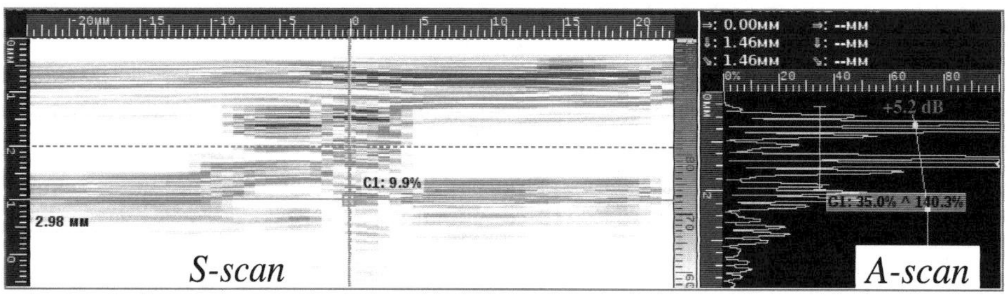

Figure 2. Example of S-scan and A-scan images.

Another widely used ultrasonic method is shadow, based on the weakening of the intensity of elastic vibrations of the ultrasonic frequency from the defective structure in the material. The control process consists in installing ultrasonic oscillation emitting and receiving transducers coaxially from different sides of the monitoring object. Oscillations, passing through the material and falling on the defect, are reflected as a result of which the intensity of the oscillations decreases. This method is well proven for the control of monolithic products. In this case, the advantage of the method is the absence of an uncontrolled zone between the transducers, as well as in the simpler configuration and the minimum necessary requirements for reference samples. However, a significant drawback of the method is the need for two-way access to the object of control.

IOP Conf. Series: Journal of Physics: Conf. Series **967** (2018) 012001 doi:10.1088/1742-6596/967/1/012001

Along with acoustic methods, the radiation method based on recording and analyzing ionizing radiation after its interaction with the object of control has become widespread [2]. In view of the characteristic features of the structure of the composite material, one of the important conditions for NDT of PCM products is the production of enlarged X-ray images. This is due to the close values of the density of the reinforcing material and binder, resulting in a low contrast of the images, thereby reducing the probability of detecting and recognizing possible defects in traditional radiography. Obtaining magnified X-ray images is possible only with the help of X-ray sources, whose focal spot sizes do not exceed 0.1 mm. Therefore, the main direction in the flaw detection of PCM products has found wide application of the method of microfocus radiography. The method makes it possible to detect heterogeneities in the form of pores, extraneous inclusions, structural inhomogeneities and breakages of reinforcing elements. Unlike traditional methods of X-ray studies, microfocus radiography has a number of important features. The first of these is the possibility of obtaining sharp enlarged X-ray images of various objects. In this case, the size of the focal spot, as well as the distance between the radiation source and the subject, significantly affect the magnitude of the geometric component, that is, the blurring of the resulting image. The second, no less important, feature of microfocus radiography is the possibility of obtaining informative X-ray images of the same objects of investigation with a lower radiation load than classical radiography [3].

However, one of the main drawbacks of the method is the complexity of its application to identify one of the most common technological defect such as delamination, in which the effect of "collapse" is rarely observed. This is due to the fact that in the delamination, the adhesive bonding between the layers of the reinforcing material is destroyed, with no air gap (figure 3). As a consequence, it is necessary to use special loading elements of the inspection object to expose the defect, which leads to complication of the control process.

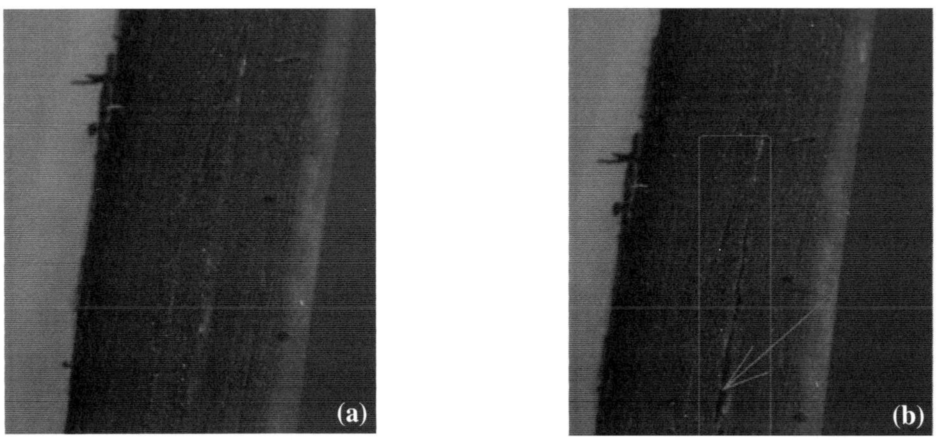

Figure 3. Microsection of the sample with destruction in the form of a delamination without load **(a)** and in a forced-loaded state **(b)**.

Along with this, in recent years in radiation flaw detection, there has been a transition to the use of digital X-ray detectors, which makes it possible to convert X-rays transmitted through the object of inspection into optical, outputted to the monitor. The merits of this method include: instantaneous acquisition of an X-ray image, input and storage of control images with further processing of results, the ability to transfer data over external and local networks. This led to the development of the method of computed tomography. This method was widely used in connection with the fact that it makes it possible to detect separately taken carbon fibers and their cliffs in carbon plastics, with a resolution of several μm.

However, like any other method, this method has its drawbacks due to both the high cost of control technology and the complexity of the application for flaw detection of large-sized products. This is due to the fact that in order to restore the volume structure of the monitoring object, it is necessary to obtain 360 to 720 pictures with a circular rotation of 360°, which is not always possible when controlling large objects.

In addition to all these methods, there is also such a promising method of NDT of PCM products as a thermal method or thermography method [4]. The method is based on measuring, monitoring and analyzing the temperature of the monitored objects. The main condition for the application of thermal control is the presence of heat flows in the controlled object. The process of transferring heat energy, releasing or absorbing heat in an object, causes its temperature to vary with respect to the environment. The temperature distribution over the surface of the object is the main parameter in the thermal method, as it carries information about the features of the heat transfer process, the mode of operation of the object, its internal structure and the presence of hidden internal defects. The heat fluxes in a controlled facility can occur for various reasons. Thermal control, in turn, is divided into two methods, active and passive.

An active method of thermal control is applied when, during operation, the object under inspection is not exposed to sufficient thermal impact, or in the case when the measurement of the temperature of the object during operation is technically impossible, such as with a helicopter blade. This method involves heating the object with special external sources of energy to create heat flows. Unlike the active method, the passive method of thermal control does not need an external source of thermal impact, the thermal field in the object of control arises when it is used or manufactured. With passive control, both the constantly acting natural thermal loading of the monitoring object and the transient thermal processes can be used. The temperature change recorded by the thermal method can reach 0.01 °C, which makes it possible to fix places of energy concentrations in which plastic deformations accompanied by heat release can occur. Based on the obtained data on the location of the energy concentrator, it becomes possible to determine the location and size of the defect. It is obvious that for defectoscopy of PCM products it is expedient to use an active thermal method that allows to detect heterogeneities in the form of delaminations in the layered structure of a composite material, as well as inhomogeneities in the form of shell exfoliations from the cellular structure in a multilayer structure.

Thus, there is a wide range of NDT methods and instruments used for flaw detection of PCM products. Such a number of methods is due to a variety of design features, geometric parameters and internal structure of inspected objects, which affects the use of one or another NDT method.

Acknowledgments

The work was supported by the Ministry of Education and Science of the Russian Federation (agreement No. 03.G25.31.0232) within the framework of the implementation of the RF Government Decision No. 218 "On Measures of State Support for the Development of Cooperation of Russian Higher Educational Institutions and Organizations Implementing Comprehensive Projects for the Creation of High-Tech Production".

References

[1] Potrakhov N N, Zhamova K K, Bessonov V B, Gryaznov A Y and Obogovsky A V 2015 *PNRPU Aerospace Engineering Bulletin* **43**

[2] Klyuev V V 2003 *Nondestructive testing* (Moscow: Mechanical engineering)

[3] Anoshkin A N, Osokin V M, Tretyakov A A, Potrakhov N N and Bessonov V B 2017 *Journal of Physics: Conference Series* **808** 012003

[4] Troitsky V A, Karmanov M N and Troitskaya N V 2014 *Technical diagnostics and non-destructive testing* **3** 29–33

A method of monitoring contact (pointed) welding

V B Bessonov, N E Staroverov, I A Larionov, K K Guk and A V Obodovskiy

Saint-Petersburg Electrotechnical University "LETI", 197376, Saint-Petersburg, Russia

E-mail: vbbessonov@yandex.ru

Abstract. The technology of welding parts of different thicknesses from various materials is improved, which is why the range of applied types and methods of welding is constantly expanding. In this regard, the issue of monitoring welded joints is particularly acute. The goal was: to develop a method of non-destructive radiographic inspection of point welds with a high accuracy rating of its quality.

With the development of technology, the technology of welding parts of different thicknesses from various materials is improved, which is why the range of applied types and methods of welding is constantly expanding. Currently, parts are welded with a thickness of several micrometers to several meters, made not only from structural steels, but also from special alloys based on non-ferrous and refractory metals, and also from composite materials [1]. In this regard, the issue of monitoring welded joints is particularly acute.

Allowing finding welding defects of macro dimensions, existing techniques are not sufficiently sensitive to microdefects, as well as to structural changes and, accordingly, changes in the mechanical properties of materials. Special difficulties raises a question of quality control at contact (pointed) welding of products [2].

In this regard, the goal was: to develop a method of non-destructive radiographic inspection of point welds with a high accuracy rating of its quality.

The most important distinguishing feature of the use of microfocus sources for radiography (the so-called microfocus shooting) is the possibility of obtaining sharp enlarged X-ray images of various objects. Depending on the specific size of the focal spot, the geometric parameters of shooting and some other factors, the magnification ratio of the size of the object in image compared to its true dimensions may range from several units to several hundreds, while maintaining quality and informative image (figure 1).

The main feature of this scheme is that the subject is positioned at a certain distance from the focal spot of the radiation source (focal length) and the entrance window of the receiver. The ratio of the indicated distances determines the image magnification ratio. Obviously, the microfocusing method of surveying makes it possible at a much higher level to assess the main criteria for the quality of a spot weld, such as:

– kernel presence;
– kernel depth;
– presence of drip fraction and caverns.

The main interest is determination of spot welding kernel depth. To form an algorithm for determining this parameter, preliminary tomographic studies were performed on a prototype of microfocus computer tomograph developed at the department of electronic instruments and devices. As objects of a research double and threefold connections from steel plates were used. The results of the volumetric reconstruction for the internal structure of the investigated objects are shown in figure 2.

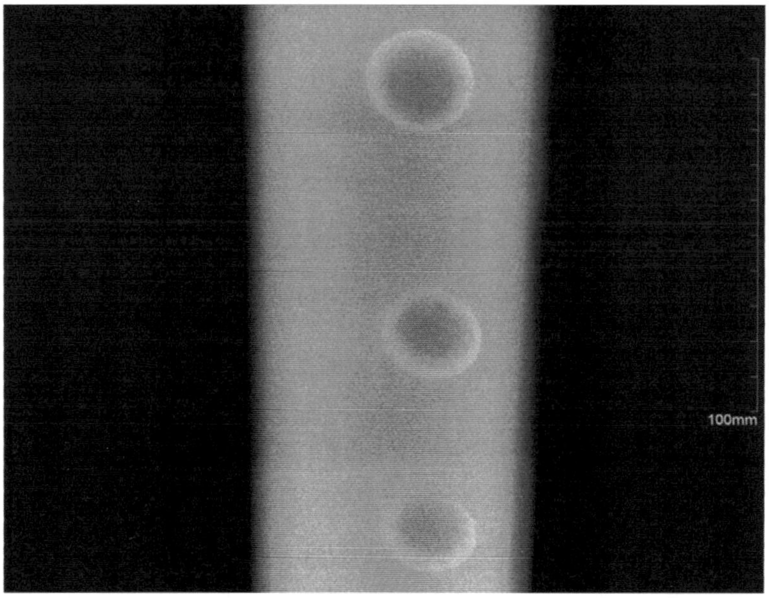

Figure 1. The image of a spot welded joint obtained by a microfocus method of shooting.

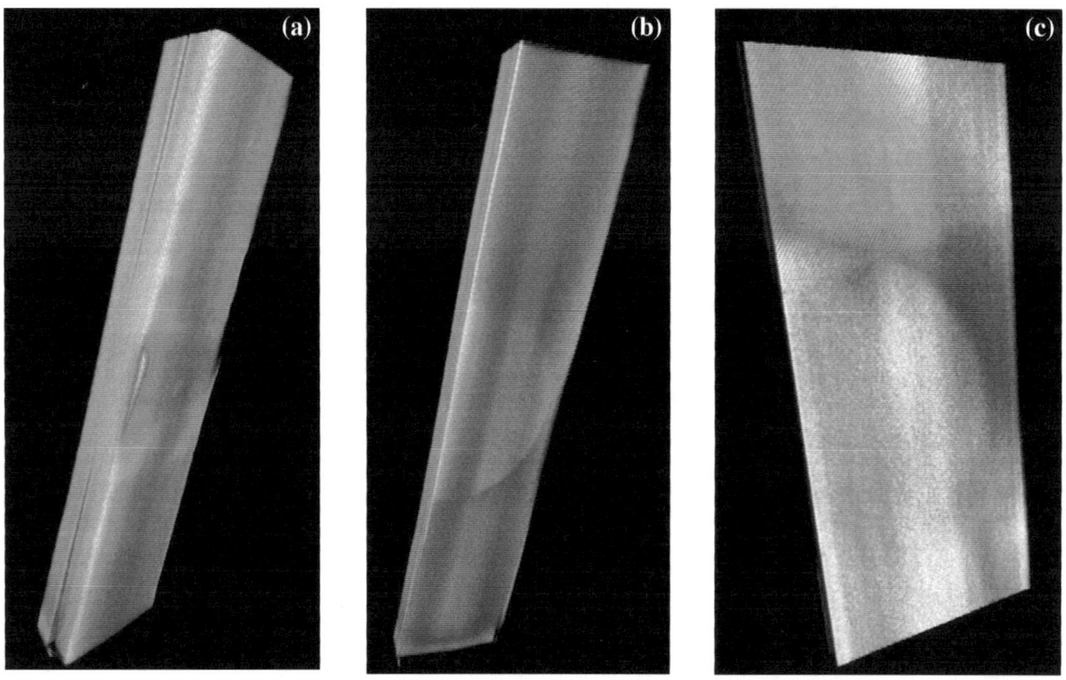

Figure 2. Tomography of samples with contact welded connections: **(a)** – three-dimensional reconstruction of two-layer sample; **(b)** – cut in the field of contact welding kernel for two-layer sample; **(c)** – cut in the field of contact welding kernel for three-layer sample.

Figures 2(b) and 2(c) clearly show an area of the kernel, thus there are all preconditions to create a method for control the kernel depth of contact (spot) welding.

To determine the depth it is necessary to capture the object from two angles: after the first (direct) image, it is necessary to move the radiation source a certain distance parallel to the plane of the sample under study (figure 3). This operation is necessary for a more accurate assessment of the welded joint quality as depending on the location of welding can significantly change the physical parameters of the resulting structure.

By measuring the image size, the distance from the axis of the radiation source to the image, the distance to the object under study and the source shift, it is possible to determine the actual size of the spot welded joint, and, as a consequence, its position in the material.

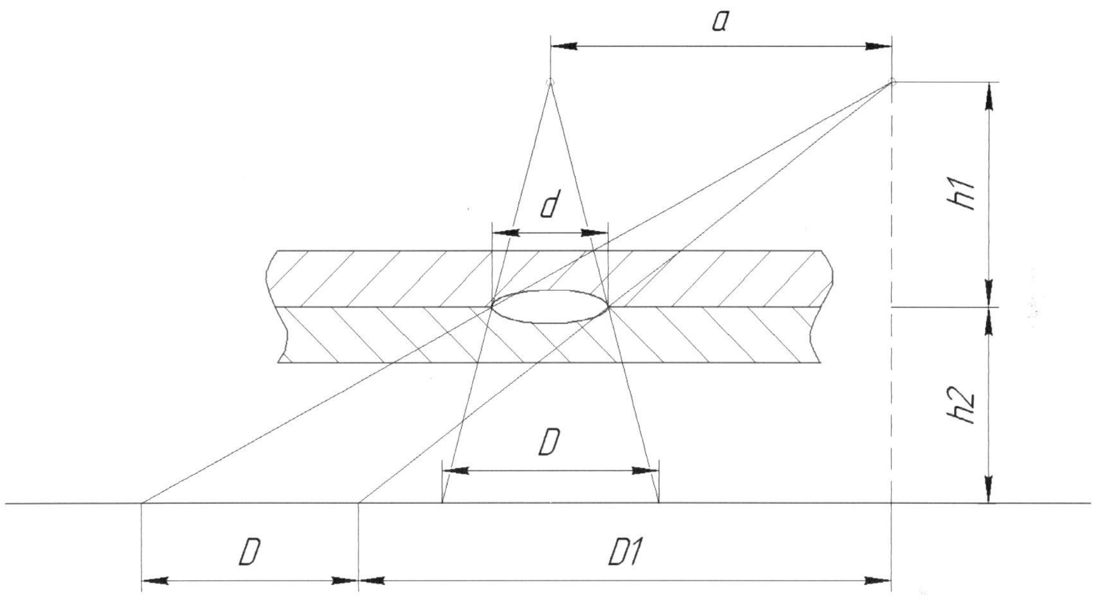

Figure 3. Scheme for obtaining the images to determine the depth of welded joints: d – actual size of the weld; D – its enlarged image; D_1 – distance from the axis of the radiation source to the image; a – shift of radiation source; h_1 – distance from the source to the center of the object under control; h_2 – distance from the center of the investigated object to the receiver.

The relationship between the true size of the object and its magnified image described by the formula:

$$D=\frac{h_1+h_2}{h_1}d\,. \tag{1}$$

Also by means of elementary mathematical transformations we can receive the following formula by analogy:

$$D+D_1=\frac{h_1+h_2}{h_1}\left(a+d/2\right). \tag{2}$$

Expressing parameter d in a formula (2) and substituting the received expression in a formula (1):

$$\frac{h_1}{h_1+h_2}=\frac{2a}{D+2D_1}\,. \tag{3}$$

Representing parameter h instead of $h1$ and $h2$ sum (distance from a source to the receiver which can be measured unambiguously), we will receive a final formula for calculation of a depth of pointed welded connection:

$$h_1 = \frac{2a \cdot h}{D + 2D_1}. \tag{4}$$

The offered control method of the products made by means of contact (pointed) welding allows to determine with high precision one of key parameters – a kernel depth.

Acknowledgments
Work is performed with financial support and at the expense of a grant of the President of the Russian Federation for the state support of young Russian scientists MK-1582.2017.8.

References

[1] Anoshkin A N, Osokin V M, Tretyakov A A, Potrakhov N N and Bessonov V B 2017 *Journal of Physics: Conference Series* **808** 012003
[2] Bessonov V B, Obodovskiy A V, Gryaznov A Y, Klonov V V, Larionov I A and Osokin V M 2017 *Journal of Physics: Conference Series* **872** 012036

IOP Conf. Series: Journal of Physics: Conf. Series **967** (2018) 012003 doi:10.1088/1742-6596/967/1/012003

About the features of the design of X-ray systems for nondestructive control in industry

V B Bessonov, A V Obodovskiy and Y N Potrakhov

Saint-Petersburg Electrotechnical University "LETI", 197376, Saint-Petersburg, Russia

E-mail: vbbessonov@yandex.ru

Abstract. X-ray methods for nondestructive control take the priority place if necessary to exercise quality control of a product, to define presence of defects and to execute expert researches. In the present work, an attempt is made to systematize some features in the design and construction of installations for X-ray non-destructive control using the example of developments conducted at St. Petersburg State Electrotechnical University "LETI".

X-ray methods for nondestructive control take the priority place if necessary to exercise quality control of a product, to define presence of defects and to execute expert researches. At the same time, a significant proportion of research is carried out in stationary conditions on the basis of specialized laboratories. Obviously, for such applications it is necessary to create appropriate conditions for work, primarily on the basis of modern safety requirements when working with ionizing radiation sources, taking into account also the nomenclature of objects that need to be studied and the requirements for research quality. The quality of the research in this case refers to the achievement of specified image characteristics, allowing to judge about the presence or absence of defects. These features primarily include:

– spatial resolution;
– contrast;
– the ratio of signal to noise;
– dynamic range [1].

In the present work, an attempt is made to systematize some features in the design and construction of installations for X-ray non-destructive control using the example of developments conducted at St. Petersburg State Electrotechnical University "LETI".

Over the past few years there have been significant changes in the sanitary rules on the territory of the Russian Federation. Significantly revised documents were issued:

– sanitary rules and regulations sanitary rules and norms 2.6.1.3289-15 "Hygienic requirements for ensuring radiation safety when dealing with sources generating X-rays at an accelerating voltage of up to 150 kV";

– sanitary rules and regulations sanitary rules and norms 2.6.1.3164-14 "Hygienic requirements on ensuring radiation safety at X-ray defectoscopy".

It is known that work with sources of ionizing radiation requires the license for the related activity which obtaining is accompanied by a number of events. However, in accordance with paragraph 1.7.2 "Basic sanitary rules of ensuring radiation safety". Exempt from control after registration of a sanitary and epidemiological certificate: electrophysical devices generating ionizing radiation, in all possible modes and conditions of operation, which is the ambient dose equivalent dose rate in any available point at a distance of 0.1 m from the outer surface of the device does not exceed 1.0 μSv/h. Under any possible operating modes product means any modes that a user can install without breaking the design or the seal manufacturer.

It should be noted that the only document currently confirming the radiation safety of the device generating ionizing radiation is the expert opinion of an accredited laboratory. In this case, the conclusion on the possibility of operating the installation without license can only be issued by the head scientific institution of Federal Service for Surveillance on Consumer Rights Protection and Human Wellbeing.

If this condition is met, obtaining permission to operate ionizing radiation sources is greatly simplified and looks as follows:

1. Submissions to Territorial Department of Federal Service for Surveillance on Consumer Rights Protection and Human Wellbeing on the territorial subject of the Russian Federation of allowing documentation (expert opinion, certificate of conformity);
2. Agreements of the application for supply with the territorial department of Federal Service for Surveillance on Consumer Rights Protection and Human Wellbeing for the subject of the Russian Federation (this item is not required in accordance with p. 3.5.1 "Basic sanitary rules of ensuring radiation safety" (however, it is recommended to perform it);
3. Training of a specialist responsible for radiation safety at the enterprise;
4. Notice Rospotrebnadzor of the implementation on the delivery the source of ionizing radiation;
5. Application and receipt of a sanitary and epidemiological certificate.

As already noted, one of the most significant requirements is the need to ensure radiation safety.

At the moment, in practice, there are three main methods for calculating protection from ionizing radiation (X-ray):

– experimental (in practice it is rarely used);
– on the basis of universal tables for calculations of protection from photon radiation of point isotropic sources;
– based on nomograms and characteristics of the X-ray tube.

The use of universal tables for the calculation of protection against photon radiation allows us to easily estimate the thickness of the necessary protective layer. Nevertheless, in view of a number of assumptions, such a calculation cannot be called accurate enough:

– the data in the tables are given for monoenergetic radiation sources;
– there is no account for secondary (scattered) radiation;
– there is no account for the characteristic radiation [2].

Despite these assumptions, apply a specified method of bringing continuous Bremsstrahlung process spectrum from X-ray tube to monoenergetic using the formula:

$$E = \frac{2}{3}eU,$$

where E – effective energy of quanta in the spectrum, keV; e – charge of electron; U – accelerating voltage, kV. Thus, for an accelerating voltage of 150 kV, the effective energy of quanta in the spectrum is 100 keV.

Knowing the radiation pattern (capacity ambient dose equivalent in different directions), it is possible, using reference tables from [2], to determine the required value of the cover thickness.

More accurate is the calculation of protection based on the calculation of nomogram parameters. Parameter of nomograms is the coefficient

$$k_1 = 2.5 \times \frac{\dot{D}_T}{\dot{D}_{T_0}} \times m \times \frac{i}{R^2} \times N,$$

where $\dot{D}_T$ – radiation output of a specific X-ray tube for which protection is determined; $\dot{D}_{T_0}$ – radiation output of some standard X-ray tube; m – the ratio of the maximum permissible design dose rate for the premises of permanent stay of personnel of category A at a 36-hour working week to design dose rate for the given conditions; i – current in the x-ray tube, mA; R – distance from the anode of the X-ray tube to the calculation point; N – directivity factor (for a direct beam $N = 1$, for scattered radiation – $N = 0.05$).

Table 1. Table for calculating protection against ionizing radiation with an average energy of $E = 100$ keV.

Rate of attenuation	Cover thickness, cm			
	Concrete	Steel	Plumbum	Tungsten
1.5	4.9	0.26	0.02	0.01
10	12.1	1.1	0.08	0.04
10^2	19.6	2.1	0.17	0.08
10^3	26.6	3.0	0.28	0.12
10^4	33.4	4.0	0.38	0.17
10^5	40.2	4.6	0.5	0.22
10^6	46.8	5.7	0.61	0.27
10^7	53.4	6.6	0.72	0.32

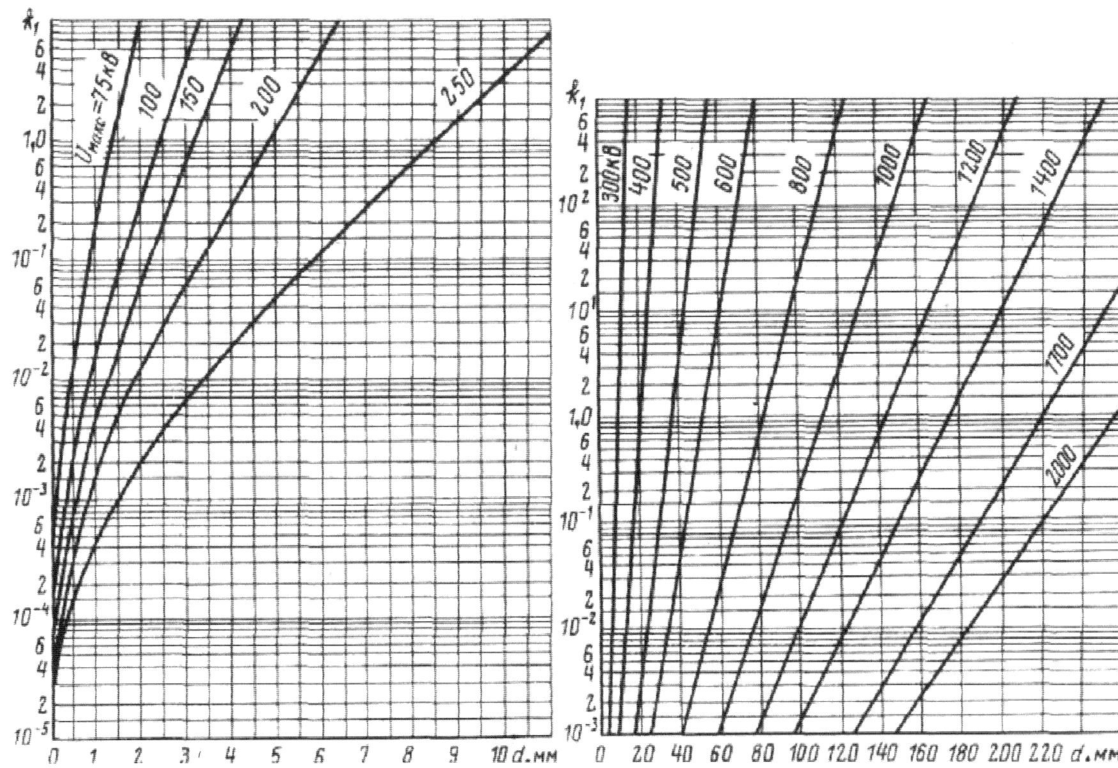

Figure 1. Nomograms for calculation of protection from plumbum.

Protection is calculated based on the nominal current and voltage values of the X-ray tube. The table of power of an exposition dose of X-ray radiation (beam return) is usually provided in technical documentation on X-ray tubes and devices under the set measurement conditions (table 2).

Table 2. Data on the dose power of X-ray radiation from a X-ray tube for the device INTROVOLT 100BE

Device type	Tube type	Exposure dose rate, R/min	Measurement conditions		
			Distance, mm	Voltage, kV	Beam current, mA
INTROVOLT 100BE	1BPV15-100	40	500	100	10

From the data presented in table 2 it is possible to determine a radiation exit of a X-ray tube by the following formula:

$$\dot{D}_T\left[\frac{\text{mGy}\times\text{m}^2}{\text{mA}\times\text{min}}\right]=\frac{\dot{X}\left[\frac{\text{R}}{\text{min}}\right]\times R^2\left[\text{m}^2\right]\times 10}{i\left[\text{mA}\right]},$$

where, according to table 2, $\dot{X}$ – exposure dose rate; R – distance; i – beam current; 10 – transition coefficient from R to Gy. Substituting these values we get:

$$\dot{D}_T=\frac{40\times 0.25\times 10}{10}=10\left[\frac{\text{mGy}\times\text{m}^2}{\text{mA}\times\text{min}}\right].$$

Next, it is necessary to determine the maximum permissible dose rate for the premises of the relevant categories of personnel (population). The permissible dose rate is determined taking into account the dose limit per year with a safety factor of 2 according to formula:

$$AD_0\left[{}^{\mu\text{Sv}}\!/\!_{\text{hour}}\right]=\frac{DL\left[{}^{\mu\text{Sv}}\!/\!_{\text{year}}\right]}{2\cdot 1700\left[{}^{\text{hour}}\!/\!_{\text{year}}\right]}.$$

The dose limit for Personnel A is 20 mSv/year, for Group B personnel – 5 mSv/year, for the population – 1 mSv/year. Then allowable dose (AD_0) for the personnel of group A is 6 mSv/h, for the personnel of group B – 1.5 mSv/h, for the population – 0.3 mSv/h.

The radiation output of a standard X-ray tube is determined graphically by reference [2].

It should be noted that in practice, a combined approach to the definition of protection is often implemented. Also effective is the collimation of the X-ray beam by lead diaphragms. With the help of this technique, it is possible to significantly reduce the mass-dimensional characteristics of the equipment by using a "two-stage" protection against ionizing radiation.

Thus, the main approaches to the organization of protection against ionizing radiation when designing installations for nondestructive control in the industry are considered.

Acknowledgments
The Russian Science Foundation have supported this work as the project entitled "The robotic system is multi-view X-ray microfocus aircraft parts and assemblies made of polymer composite materials with complex action". Project number 16-19-00155.

References

[1] Potrakhov N N, Gryaznov A Yu, Potrakhov Ye N and Bessonov V B 2015 *Vakuumnaya tekhnika i tekhnologiya* **2** 153
[2] Mashkovich P and Kudryavtseva A V 1995 *Zashchita ot ioniziruyushchikh izlucheniy* (Moscow: Energoatomizdat)

Production of high-performance three-layer emitters of X-ray tube cathodes by diffusion welding

N A Golovin[1], M L Taubin[1,2], D A Chesnokov[1] and D S Kiselev[1,2]

[1] FSUE "SRI SIA "LUCH", Podolsk, Moscow Region
[2] NRNU "MEPhI", Moscow

Abstract. The paper presents the description of the investigation of x-ray tube cathode emitters made of refractory metals by diffusion welding. The weld seam structure has been analyzed. Comparison of the emissivity characteristics of three-layer Ta–W–Ta and Nb–W–Nb cathode emitters has been made.

1. Introduction

Development of modern engineering brings forth the necessity to improve the information capacity of X-ray devices such as the angiograph and particularly – tomographic scanner, which is highly dependent on the radiation intensity. One of the ways to enhance the radiation intensity is to use low work function materials for cathode emitters.

As it is commonly known, the dependence between the emission current density and the work function is defined by Richardson's equation (1):

$$j = A \cdot T^2 \cdot e^{-\varphi/kT}, \qquad (1)$$

where j – current density; T – temperature in Kelvins; φ – work function; k – Boltzmann constant. Since the operational temperatures of X-ray tube cathodes are usually 2000 °C or more, the most promising way to increase the current density is to reduce the work function [1].

According to the literature references, the work function of tungsten commonly used in cathodes is 4.55 eV, that of tantalum is – 4.11 eV, and niobium – approximately 3.99 eV. However, in their pure form these materials are not well suited for cathode emitters because of lower melting point as compared to tungsten and worse high-temperature creep resistance as well [2].

Therefore, the authors herein propose a three-layer design for a flat cathode with tungsten substrate and niobium or tantalum emitter. The authors suppose that the optimal method of production of such an assembly is diffusion welding of foils.

2. Experimental

The welding was performed using a special vacuum facility. A high-temperature component made of refractory metal was used as the heating source. Figure 1 presents the schematics of the diffusion wielding. Metal foils with thicknesses of approximately 100 µm for tungsten and 50 µm for niobium and tantalum were used as the samples. The samples were pre-treated: etched with acidic solutions and vacuum annealed at T = 1000 °C in order to remove the contaminants and oxides from the surface. The welding was performed within the temperature range 1600–2000 °C and under pressure of approximately 10–20 MPa. Duration of exposure was 15 min. The resulting plates were cut into 3×18 mm stripes. Some samples were used for performance of metallographic studies to control the weld seam quality, and the remaining samples were mounted inside the special chamber to measure the work function (figure 2).

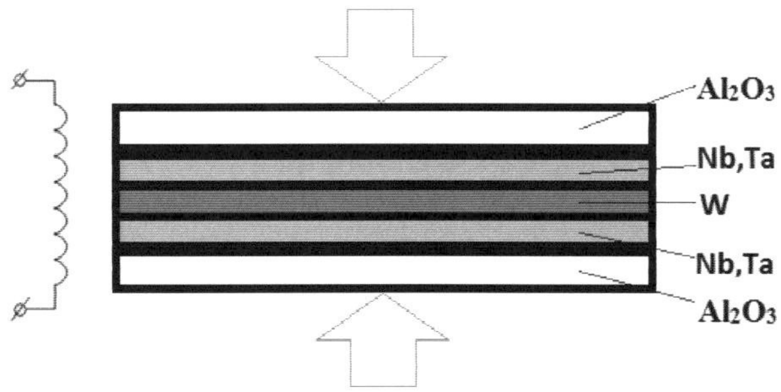

Figure 1. Schematics of the diffusion welding.

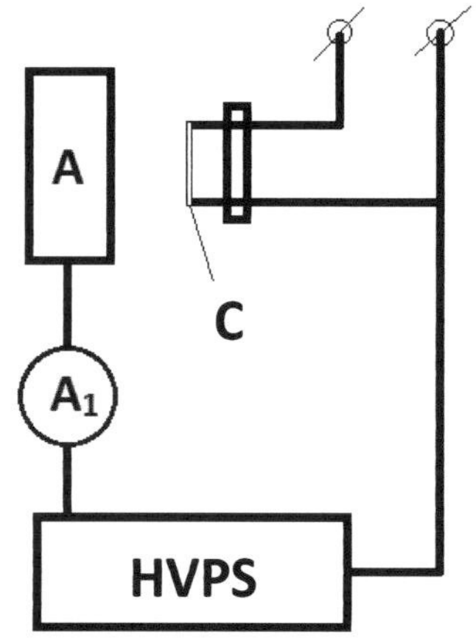

Figure 2. Schematics of the work function measuring apparatus.

The anode (A) can be moved in vertical direction to measure the cathode temperature with a pyrometer. The high-voltage power source (HVPS) is operational with voltage of up to 1000 V. The emission current (I) is measured with the amperemeter (A_1) and the work function is then calculated using the Richardson equation (1).

3. Results and discussion

Metallographic studies of the samples of three-layer Nb–W–Nb and Ta–W–Ta cathodes made by diffusion welding have been conducted. Figure 3 presents the cross-section images of these samples.

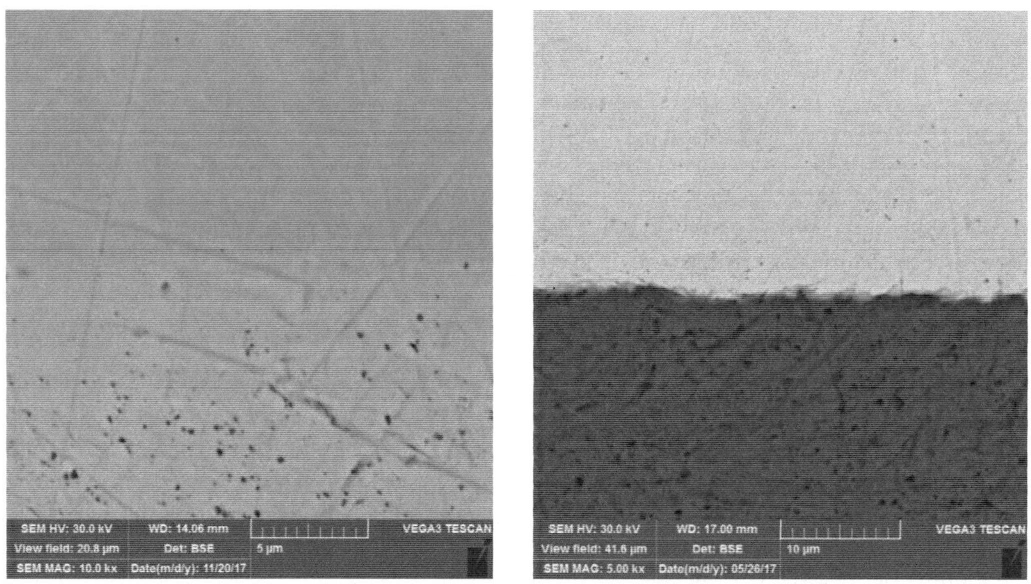

Figure 3. Weld seam of the samples made by diffusion welding: **(a)** – Ta–W–Ta; **(b)** – Nb–W–Nb.

Figure 3 clearly shows that diffusion welding ensures the high-quality connection between the two materials. Lack of foreign particles in the weld seam is indicative of decent preparation treatment of the welded surfaces.

IOP Conf. Series: Journal of Physics: Conf. Series **967** (2018) 012004 doi:10.1088/1742-6596/967/1/012004

On completion of the metallographic studies, multilayer cathode emitters were made by erosive cutting and work function was measured. Results of these studies are presented in figures 4 and 5.

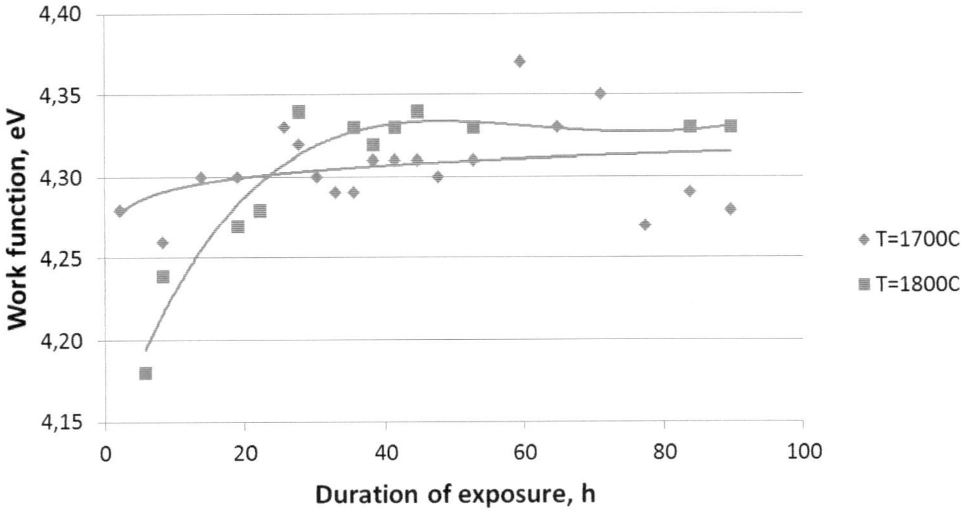

Figure 4. Work function of the three-layer Nb–W–Nb cathode.

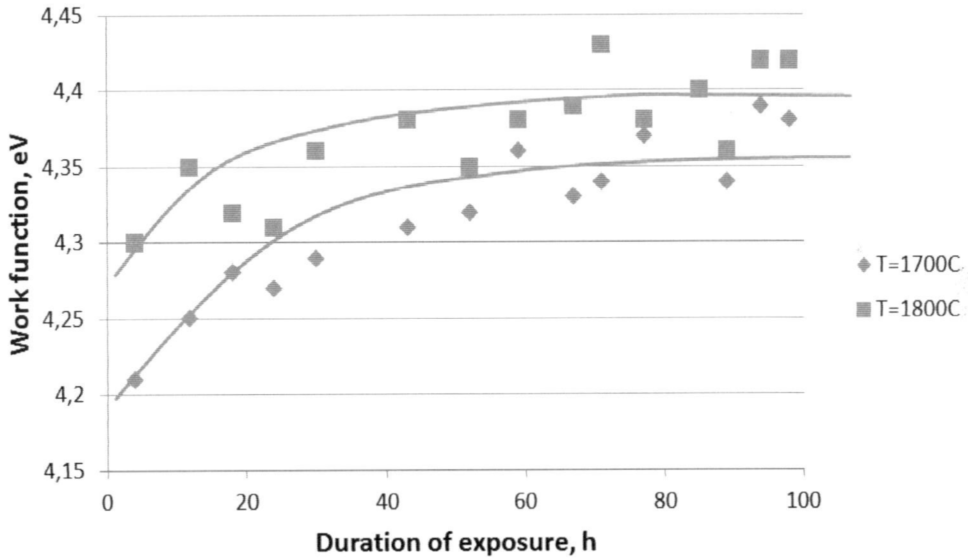

Figure 5. Work function of the tree-layer Ta–W–Ta cathode.

As it can be seen from the plots, the work function increases with the duration of exposure, becoming constant after 40–50 h. It is also worth noting that the work function is slightly less than that of niobium or tantalum. Probably, it is caused tungsten diffusion into tantalum and niobium.

4. Conclusions

1. Three-layer Nb–W–Nb and Ta–W–Ta cathode emitters were made by the method of diffusion welding.
2. The modes of welding and methods of sample preparation used in this work were proved to assure the proper connection between the metal foils.

IOP Conf. Series: Journal of Physics: Conf. Series **967** (2018) 012004 doi:10.1088/1742-6596/967/1/012004

3. Life testing has shown that work function of such emitters is lower than that of tungsten, which should make it possible to increase the x-ray tube emissive current.

References

[1] Alekseev S V, Taubin M L and Yaskolko A A 2014 *Nanocomposits in X-ray technologies* (Moscow: Technosphera)
[2] Gontar A S, Konoplev E E and Taubin M L 2013 *Medical Physics* **4**

Microfocus X-ray seeds separator

A Yu Gryaznov, N E Staroverov, E D Kholopova and K K Guk

Department of electronic instruments and devices, Saint Petersburg Electrotechnical University "LETI", 197376, Saint Petersburg, Russia

E-mail: nik0205st@mail.ru

Abstract. In this paper the model of X-ray seeds separator is described. The main parameters of the seed separator are determined. The operating modes of the X-ray apparatus were determined; with these modes the greatest contrast of the image was achieved. For the research, seeds of almost 40 different plant species were selected. For each type of seed, the optimal parameters of the imaging were calculated; also during the experimental studies the parameters were checked.

The production of high quality seeds is an important task of developing agro-industrial complexes. Nowadays X-ray methods of nondestructive testing of seeds quality begin widely using. The main advantages of these methods are the high speed of analysis, the possibility of obtaining a large amount of information and the high probability of recognition of defects and pests.

Today there are two main methods of X-ray filming: the contact method and the method with a magnification of the image [1, 2]. To apply the first (contact) method the radiation source with the large focal spot (focal spot size more than 0.1 mm) is used. The filming object is located at a sufficiently large distance from the radiation source and as close as possible to the X-ray image detector. To apply the second (with magnification) method the microfocus radiation source (focal spot size less than 0.1 mm) is used. This method is called microfocus radiography. The filming object is located at the specified distance between the radiation source and the image detector and the ratio of this distance determinates the image magnification of the object.

Microfocus radiography allows getting an enlarged image of the object. That is why this method is preferable for the radiography of seeds. This method allows getting sharp images of seeds with high magnification, which is not possible using contact method. Thus, the method of microfocus radiography can reliably detect the variety of seed defects, including the percentage of damage, empty seeds and pests [3].

For the separation of qualitative and substandard seeds, an X-ray seed separator model was proposed. In this model, X-ray radiation is used to obtain a shadow image of a group of seeds, after which digital processing of the image is performed. Based on the received data, the decision is made whether the seed is damaged or not, and, consequently, the damaged seeds are separated from the intact ones.

The model of the X-ray seed separator includes an X-ray protection chamber 1, a microfocus X-ray source 2, radiation source displacement system 3, X-ray detector 4, consisting of a scintillator 5 and a matrix of photosensitive transistors 6, seed displacement system 7, magnetic focusing system 8,processing and control unit 9. The processing and control unit is connected to the X-ray radiation detector, the X-ray radiation source and the magnetic focusing system. A schematic representation of the model is shown in figure 1. The model is controlled by the operator, which loads the research objects into the seed displacement system, and also controls the operation of the separator.

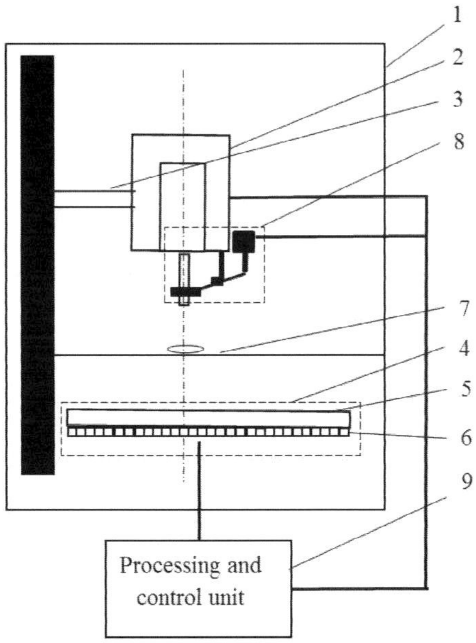

Figure 1. Schematic of the model of seeds separator.

The coefficient of image magnification is determined when the operator selects the type of seeds to be examined. The coefficient is calculated on the basis of the minimum size of the defects that can be presented in these seeds, according to the following formulas:

$$A = a \cdot \frac{f_2}{f_1} \quad (1)$$

$$H = d \cdot \frac{f_2 - f_1}{f_1} \quad (2)$$

where a – size of the defect; A – size of the defect image; H – blur of the defect image; d – focal spot size of microfocus X-ray radiation source; $f1$ – distance from the focal spot to the seed; $f2$ – distance from the microfocus X-ray radiation source to X-ray detector. The separator works as follows: the operator sets the type of seeds to be examined: the voltage and current of the X-ray tube are automatically set, depending on the type of seeds. Also, depending on the type of seeds, the distance from the source of radiation to the object of research varies, thus changing the image magnification factor. The X-ray radiation, generated by the microfocus source passes through the seeds and is unevenly attenuated by the elements of seeds structure. X-ray image the image is recorded and converted into a digital electrical signal by an X-ray detector. The processing and control unit uses special software to process the digital image of each seed and determine the quality of all seeds.

To determine the quality of the seeds, a bank of standard X-ray images of seeds was collected, which were obtained earlier in the X-ray studies from different seeds cultures. During the development of the method, the operating modes of the X-ray apparatus were determined; with these modes the greatest contrast of the image was achieved. Samples of seeds for research were provided by the Botanical Garden of Peter the Great of the BIN RAS (St. Petersburg). For the research, seeds of almost 40 different plant species were selected. To carry out the research, a microfocus radiation source RAP50 is used, which having a range of variation of the anode voltage of the X-ray tube 5...50 kV and the range of the anode current variation of 20...200 μA.

For each type of seed, the optimal parameters of the imaging were calculated; also during the experimental studies the parameters were checked. At X-ray inspection of seeds research objects are usually located on cards, in strict order. The card has a working field of X-ray transparent material with an area that allows the total number of seeds to be analyzed.

To obtain a bank of seed images, cardboard cards were also used. However, since when sorting the seeds with the help of a separator they will not be located separately from each other, in strict order, the experiments were carried out which made it possible to find out that all hidden defects are reliably recognized in chaotically located seeds.

As a result of the research, a bank of X-ray images of seeds of various plant species, representatives of different families was assembled. The main parameters of the developed X-ray seed separator are determined. It is shown that in the separator it should be possible to change the voltage of the X-ray tube from 14 to 30 kV, tube current from 60 to 200 μA over the operating voltage range. The image magnification ratio should vary from 3 to 20 times. For X-ray separation of seeds, microfocus sources of X-ray radiation should be used. The use of such radiation sources makes it possible to obtain images with high magnification. The proposed technique for automatic X-ray seed separation makes it possible to simplify the work and reduce labor costs for cleaning seeds to obtain results promptly and without seed quality testing by traditional methods.

References

[1] Gryaznov A Yu, Potrakhov E N and Potrakhov N N 2008 *Petersburg Journal of Electronics* **2–3** 163–6

[2] Alpatova V G, Vasilyev A Yu, Kiselnikova L P, Potrakhov N N, Gryaznov A Yu, Selyagina A S and Petrovskaya V V 2010 *Institute of Dentistry* **46** 80–1

[3] Tkachenko K G, Kapelyan A I, Gryaznov A Yu and Staroverov N E 2015 *Bulletin of the Botanical Garden Institute FEB RAS* **13** 34–40

IOP Conf. Series: Journal of Physics: Conf. Series **967** (2018) 012006 doi:10.1088/1742-6596/967/1/012006

Evaluation of low-dose CT implementation for lung cancer screening in a general practice hospital

D V Karostik[1,2], I G Kamyshanskaya[2,3], V M Cheremisin[2,3], A A Drozdov[2,3] and A V Vodovatov[4]

[1] Saint-Petersburg State Research Institute of Phthisiopulmonology of the Ministry of Healthcare of the Russian Federation, Ligovsky Pr., 2-4, 191036, St. Petersburg, Russia
[2] FGBOU VO "Saint-Petersburg state University", Department of Oncology, 21 line V. O., 8A, 199034, St. Petersburg, Russia
[3] SPb GBUZ "Urban Mariinsky hospital", Liteiniy pr., 56, 191014, St. Petersburg, Russia
[4] St. Petersburg Research Institute of Radiation Hygiene n.a. Professor P.V. Ramzaev, Mira st. 8, 197101, St. Petersburg, Russia

E-mail: d.karostik@gmail.com

Abstract. The aim of the current study was to evaluate the possibility of the implementation of LDCT for the screening for lung cancer and tuberculosis in a typical general hospital practice. Diagnostic and economic effectiveness, patient doses and the corresponding radiation risks for LDCT were compared with the existing digital chest screening radiography. The results of the study indicate that the implementation of LDCT allowed verifying false-positive cases or providing additional excessive diagnostic information, but did not significantly improve the sensitivity of screening. Per capita costs for LDCT were higher compared to digital radiography up to a factor of 12; corresponding radiation risk – by a factor of 4. Hence, it was considered unjustified to implement LDCT in a general practice hospital.

1. Introduction

Lung cancer is the second leading cause of death in the Russian Federation after blood circulatory system diseases, corresponding to about 10 % in the structure of the primary malignant cancer incidence. Lung cancer corresponds to 28 % and 7 % in the structure of the cancer mortality for males and females, respectively. One of the reasons for that is the ineffective diagnostic of lung cancer on early stages [1]. The detection of lung cancer is of paramount importance in mortality reduction. Recent developments in computed tomography (CT) allow using low-dose CT (LDCT) for the lung cancer screening. LDCT protocols use lower tube voltage and tube current as well as higher slice thickness, compared to standard chest CT protocols. That allows keeping the typical effective dose per LDCT examination in a range of 1 mSv, preserving the acceptable diagnostic image quality for the lung nodules detection. Clinical trials in different countries indicated a significant increase in the number of the early lung cancers detected with LDCT compared to the conventional chest screening methods [2–4]. These trials were focused on different risk groups, commonly considering the age (50–80 years) and the smoking experience (20+ pack-years) [5].

However, chest screening in the Russian Federation is mainly focused on the detection of tuberculosis, professional and inflammatory diseases of the chest. It is performed using conventional chest radiography on the dedicated digital or film-based X-ray units (film-based and digital fluorography). Almost all the working Russian population undergoes annual conventional X-ray screening on an outpatient basis. Since 2014, there is an ongoing discussion on the implementation of LDCT in the screening programs, but the number of studies and trials is limited.

Hence, the aim of the current study was to evaluate the possibilities of the implementation of LDCT for the lung cancer screening in a typical general practice hospital in Russia. Diagnostic and economic effectiveness, patient doses and the corresponding radiation risks for LDCT were compared with the existing digital chest screening radiography.

2. Materials and methods

The study was performed in the X-ray and CT departments in a general practice St. Petersburg "Urban Mariinsky hospital" in 2016–17. The study consisted of two parts: retrospective and prospective. A retrospective part was focused on the analysis of the sample of 8554 patients that underwent a conventional digital X-ray chest screening in 2016. The following parameters were considered: age and gender of the patient; parameters of the chest X-ray examination, relevant for the estimation of the effective dose and the diagnostic conclusion of the radiologists. Data was obtained from the patient medical cards and DICOM archives with subsequent anonymization.

Prior to the study, criteria for the selection of the patients corresponding to the risk groups were developed based on the data from the foreign trails. Criteria included age of 40+ years and smoking experience of 20+ pack-years. Patients corresponding to the criteria were selected from the patient flow and invited to participate in the study with the signed patient informed consent. A total of 126 patients agreed to participate in the prospective study. They were divided into two sample based on the result of the conventional X-ray chest screening: 103 patients without pathologies and 23 patients with any identified pathology. 10 random patients from a sample without pathologies and all patients with pathologies were referred to LDCT (a total of 33 patients) with a subsequent chest CT. The design of the study was approved by the local ethical committee.

Conventional chest X-ray screening was performed on a digital X-ray unit FC-01 (ZAO "NIPK "ELECTRON", Russia) with a CCD-matrix detector, using standard chest protocol: tube voltage 100 kV, focal image distance 150 cm, field size of 35×35 cm, total filtration thickness of 5 mm Al, with grid and automatic exposure control. LDCT and CT were performed on a Somatom Definition 64 (Siemens, Germany) unit. Data on the chest LDCT and CT protocols is presented in table 1.

Table 1. Low-dose and standard chest CT protocols.

Protocol	Tube voltage, kV	Tube current, mA	Pitch	Collimation	Slice thickness, mm
LDCT	100	40	1.35	64×0.6	5.0
CT	120	80–200[1]	1.35	64×0.6	1.0

[1] with automatic mA modulation

Conventional chest X-ray images were read by the radiologist with 4 years of experience; LDCT and CT images – by the head of the CT department with 10 years of experience.

Effective doses for the conventional chest X-ray screening were estimated using PCXMC 2.0 software (STUK, Finland) based on tissue weighting coefficients from ICRP Publication 60; effective doses for the CT examinations were estimated using conversion coefficients from the existing Russian guidelines [6]. Radiation risks were estimated according to the guidelines [7].

Diagnostic effectiveness of the conventional chest X-ray screening and LDCT was evaluated based on the comparison of sensitivity, specificity, negative and positive predictive values. Economic effectiveness was evaluated based on the comparison of the per patient expenses during a 10-year period (minimal lifespan for the X-ray equipment before the replacement). The following factors were considered: staff yearly salary; price of the new equipment; technical maintenance and service expenses. The prices are provided in Russian rubles (RUR); prices are valid for 2017.

3. Results

The results of the retrospective analysis of the patient data are presented in tables 2 and 3.

Table 2. Distribution of the conclusions for the 2016 patient sample.

Result	Number of patients	%
No pathology	7339	85.79
Infiltrate	256	2.99
Consolidation	65	0.76
Single pulmonary nodule	39	0.46
Posttuberculosis calcificate	102	1.19
Disseminated processes	17	0.20
Tuberculosis	16	0.19
Malignant lesion (lung cancer/metastases)	21	0.24
Other	699	8.17

Table 3. Age distribution for the 2016 patient sample.

Age group	15–25	26–35	36–45	46–55	56–65	66–75	76–85	86–95	96+
Number of patients	939	1811	1410	1390	1433	864	579	124	4
%	10.9	21.2	16.5	16.2	16.8	10.1	6.8	1.4	0.1

Comparison of the patient doses from conventional chest X-ray screening, LDCT and chest CT is presented in table 4. Data is presented as mean ± SD (min–max).

Table 4. Comparison of the patient doses for the 2017 patient sample.

Examination	Dose-area product, $\mu Gy\cdot m^2$	DLP, mGy·cm	Effective dose, mSv
Conventional chest X-ray	71.4 ± 31.8[1] (37.4–157.2)	–	0.13 ± 0.06 (0.06–0.28)
LDCT	–	56.7 ± 5.3 (49.6–68)	0.96 ± 0.09 (0.84–1.16)
Chest CT	–	283.6 ± 26.3 (248–340)	4.82 ± 0.45 (4.22–5.78)

Comparison of the diagnostic effectiveness between the conventional chest X-ray screening and LDCT is presented in table 5.

Table 5. Comparison of the diagnostic effectiveness between the conventional chest X-ray screening and LDCT for the 2017 patient sample.

	Sensitivity, %	Specificity, %	Positive predictive value, %	Negative predictive value, %	Diagnostic effectiveness, %
Conventional chest X-ray	93	47	57	90	67
LDCT	93	100	100	95	97

Comparison of the corresponding radiation risks for the conventional X-ray chest screening and LDCT is presented in table 6. Patient age distributions are taken from table 3; effective dose values – from table 4.

Table 6. Comparison of the radiation risks between the conventional chest X-ray screening and LDCT.

Examination	Age group	Number of patients	Age-specific coefficient	Total effective dose	Collective risk	Lifetime attributable risk
Conventional chest X-ray	<18	52	2.3	7.73	0.001	0.099
	18–65	6923	0.9	1840.27	0.094	
	65+	1423	0.1	645.17	0.004	
LDCT	<18	52	2.3	49.92	0.007	0.355
	18–65	6923	0.9	6646.08	0.341	
	65+	1423	0.1	1366.08	0.008	

Comparison of the expenses per patient for the conventional chest X-ray screening and LDCT for the 10-year period is presented in table 7. Data is presented in RUR. Prices are valid for 2017.

Table 7. Structure of expenses for the conventional chest X-ray screening and LDCT.

Expenses	Conventional chest X-ray, RUR	LDCT, RUR
Staff salary		
	4800000	4800000
Maintenance		
Costs for the new X-ray/CT unit	5203175	51220000
Yearly technical maintenance	350000	2000000
X-ray tube replacement	–	5200000
Maximum number of patients per day	34	25
Number of patients for the 10-year period	85000	61750
Total expenses for the X-ray unit/CT for the 10-year period	13503175	120908000
Expenses per 1 patient	159	1958

4. Discussion

As it is visible from table 2, more than 85 % of the patients had no identified pathology according to the conclusion. The number of malignant lesions was low – only 21 cases, corresponding to 0.24 %. It can be explained by low malignant morbidity for the patient sample or by low detectability of the malignant lesions using conventional methods (digital radiography). It should be noted that age distribution for the patient sample is shifted towards age 40+ and is comparable with the patient samples used in other LDCT screening trials. Unfortunately, no data on smoking experience was available for the 2016 patient sample.

Comparison of the diagnostic effectiveness between the conventional chest X-ray screening and LDCT was performed based on the chest CT evaluation and conclusions. According to table 5, both modalities had identical sensitivity. However, the specificity and positive predictive value of the conventional chest X-ray screening were lower by a factor of 2; resulting diagnostic effectiveness – by a factor of 1.5. High sensitivity of conventional chest X-ray screening justifies its use as a screening tool – the ability to detect the pathology, with the subsequent CT or LDCT verification of the findings. According to the analysis of the radiologists' conclusions, LDCT results did not differ significantly from the conventional chest X-ray screening, providing additional or excessive information. It should be noted that the current study was performed on a limited scale due to the high patient flow in a CT department. That could definitely influence the results. For instance, there was only one case of lung cancer identified in 2017 patient sample (later verified by the biopsy). Nevertheless, the design of the study can be used later on a larger scale.

According to the data from table 6, the replacement of conventional X-ray screening by the LDCT would lead to an increase in the number of radiation-induced cancers by a factor of 4: from 1.05 cases to 4.2 cases per 100 000 patients in addition to the background cancer incidence. However, that increase is neglible in comparison with data from table 2 (246 cases of malignant lesions recalculated per 100 000 patients).

The results of comparison of economic efficiency indicate significant differences between the conventional X-ray screening and LDCT: overall expenses per patients differ by a factor of 12. Technical maintenance costs for a CT unit exceed the costs for an X-ray unit by a factor of 6, which is significant in the current Russian clinical environment. It should be noted that the another limiting factor for the implementation of the LDCT screening is the capacity of the CT-unit (number of patients per day in less compared with the X-ray unit by a factor of 1.5). The implementation of LDCT screening would lead to the installation of an additional (or several, to maintain the patient floe) CT-unit; conventional chest X-ray screening can be performed on virtually any X-ray unit. The economic factors are the main limiting factors for the implementation of LDCT for lung cancer screening as all the screening examinations are funded by the Ministry of Healthcare.

5. Conclusions

The results of the current study indicate the lack of significant differences between conventional chest X-ray screening and LDCT in terms of sensitivity, justifying the use the existing chest screening method. LDCT allowed verifying false-positive cases and acquiring additional diagnostic information. The increase in the patient effective doses and the corresponding radiation risk for LDCT was not considered to be significant. However, a significant disadvantage of LDCT as a screening tool is associated with expenses per patient and limited availability of the CT-units. Hence, the use of LDCT as a screening tool in a general practice hospital is unjustified.

References

[1] Venkatesh A, Mannudeep K, et al. 2015 *J. Thorac. Imaging* **30** 283–9
[2] Infante M, Cavuto S, Lutman F R, et al. 2009 *Am J Respir Crit Care Med* **180** 445–53
[3] Pastorino U, Rossi M, Rosato V, et al. 2012 *Eur J Cancer Prev* **21** 308–15
[4] Aberle D R, Adams A M, Berg C D, et al. 2013 *N Engl J Med* **368** 1980–91
[5] Kanodra N M, Silvestri G A and Tanner N T 2015 *Cancer* **1** 1347–56
[6] *Assessment of effective dose to the patients undergoing X-ray examinations. Methodical guidance 2.6.1.2944-11* (Moscow: Rospotrebnadzor)
[7] *Assessment of radiation risks of patients undergoing diagnostic examinations with the use of ionizing radiation. Methodical recommendations 2.6.1.0098-15* (Moscow: Rospotrebnadzor)

Operational control of large-sized polymer composite units of U-profile circular beam type by X-ray method

B V Kavalerov[1], A N Anoshkin[1], V M Osokin[1], A A Tretyakov[1], N N Potrakhov[2], V B Bessonov[2] and A V Obodovskiy[2]

[1] Perm national research polytechnic university, 614000, Perm, Russia
[2] Saint Petersburg Electrotechnical University "LETI", St. Petersburg, Russia

E-mail: osokin-kt@pstu.ru

Abstract. The advantages of using the method of microfocus radiography in the non-destructive testing of aviation products are considered in the paper, using the example of a circular beam of a U-shaped profile made of polymer composite materials. The basic types of characteristic defects of parts arising in such a type are described both in the process of their manufacture and in the process of their exploitation, namely interlayer delaminations, pores and folds. Peculiarities of obtaining pseudo-volumetric images, which allow to increase the informativity about the structure of the object of control, as well as to identify the arising heterogeneities are given. A model of a robotic system is described that makes it possible to realize a small or multi-angle survey scheme, and, in particular, to carry out tomographic studies.

At present, the use of polymer composite materials (PCM) in the aviation industry is becoming increasingly popular when developing new and improving the existing parts and assemblies. The competitive advantage of such parts in comparison with metal analogs is their smaller weight, with the same rigidity and strength characteristics, which leads to an increase in fuel efficiency, and in some cases, an increase in manufacturability and a reduction in the cost of manufacturing them. In this case, products made of PCM, have significant dimensions reaching 2 meters in diameter, subject to a wide range of static, dynamic and cyclic loads for a long period of operation. One of the characteristic examples of such a product is the circular beam of the U-shaped profile. The segment of the ring beam after the mechanical tests in the initial and forced-loaded state is shown in figure 1. In such designs, the main types of defects that occur both during the manufacture of parts and during their operation are interlayer delaminations, pores and folds, to a greater extent, formed at the places of inflection of the layers [1, 2]. For their detection, various methods of nondestructive testing are widely used. However, the detection of such defects is a complex task, since when controlling such defects, the effect of "collapse" is observed, i.e. no disclosure. Accordingly, the task of express-assessment of the quality of PCM parts and assemblies during their serial production and in the process of operation becomes paramount.

At present, one of the most informative and accurate methods of nondestructive testing for detecting typical defects is the method of microfocus radiography used in this work. Microfocus radiography is a method of obtaining X-ray images of various objects with the help of X-ray sources, the dimensions of the focal spot of which do not exceed 0.1 mm. Features of the method are widely described in the work of the author's team [3]. To solve the problem of operative highly informative control of large-sized products made of PCM circular beam of the U-shaped profile type, a microfocus method of obtaining X-ray images was applied. X-ray studies of this type of structure are advisable to perform when surveying controlled areas from a small focal length using a widely divergent beam of X-rays. To obtain information on the distribution of inhomogeneities in the internal structure of research objects, it is necessary to perform low or multi-angle survey [4].

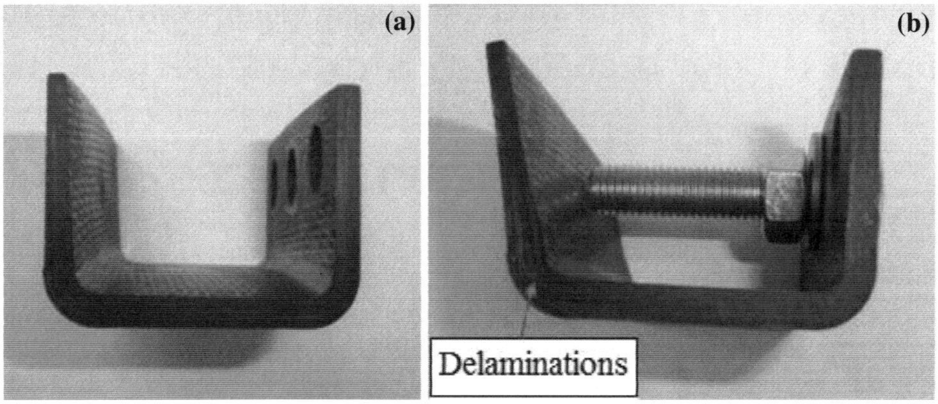

Figure 1. Composite segment of a circular beam of a U-shaped profile after mechanical tests in the initial **(a)** and forced-loaded state **(b)**.

The first method is convenient to implement, using the laminography. When it is realized, the radiation source and the object of investigation move linearly in mutually opposite directions to the maximum possible distance. This distance is determined by the geometry of the research object, the uniformity of the irradiation field, the width of the dynamic range of the visualization system, the physical dimensions of the machine for performing the scan. The second method allows a complete reconstruction of the object under study, but requires a set of projection data obtained at an angle of at least 180 degrees to the axis passing through the object of the inspection.

The implementation of these mechanisms dictates the need to create a specialized system that allows research in a similar way. Similar solutions exist in Western countries, but their cost is extremely high. In Russia, the task is solved by a team of researchers and developers from the Perm National Research Polytechnic University and the St. Petersburg State Electrotechnical University "LETI" (figure 2).

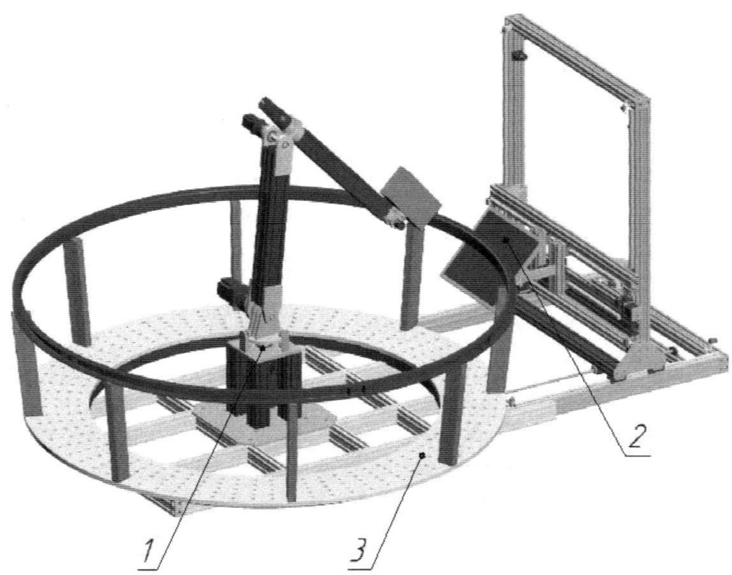

Figure 2. Model of the robotic system.

This system includes several key components:

1. Robot manipulator with a microfocus source of X-rays installed on it. The manipulator must provide irradiation of the investigated object from different angles with observance of a given accuracy of positioning and orientation with respect to the radiation receiver;
2. Frame construction with guides, providing positioning of the digital X-ray detector. Given the dimensions of the parts being studied, several images may be required to build projections from one angle;
3. Rotary table, which fixes the object under inspection.

On the basis of the projection data obtained on the layout of the machine, by the method of reverse projection with filtration and convolution, a layer-by-layer reconstruction of the internal structure of the object of investigation was performed, using 180 degrees to the axis passing through the object of control. After this operation, for the convenience of presenting the results of the study, a three-dimensional reconstruction was performed (figure 3).

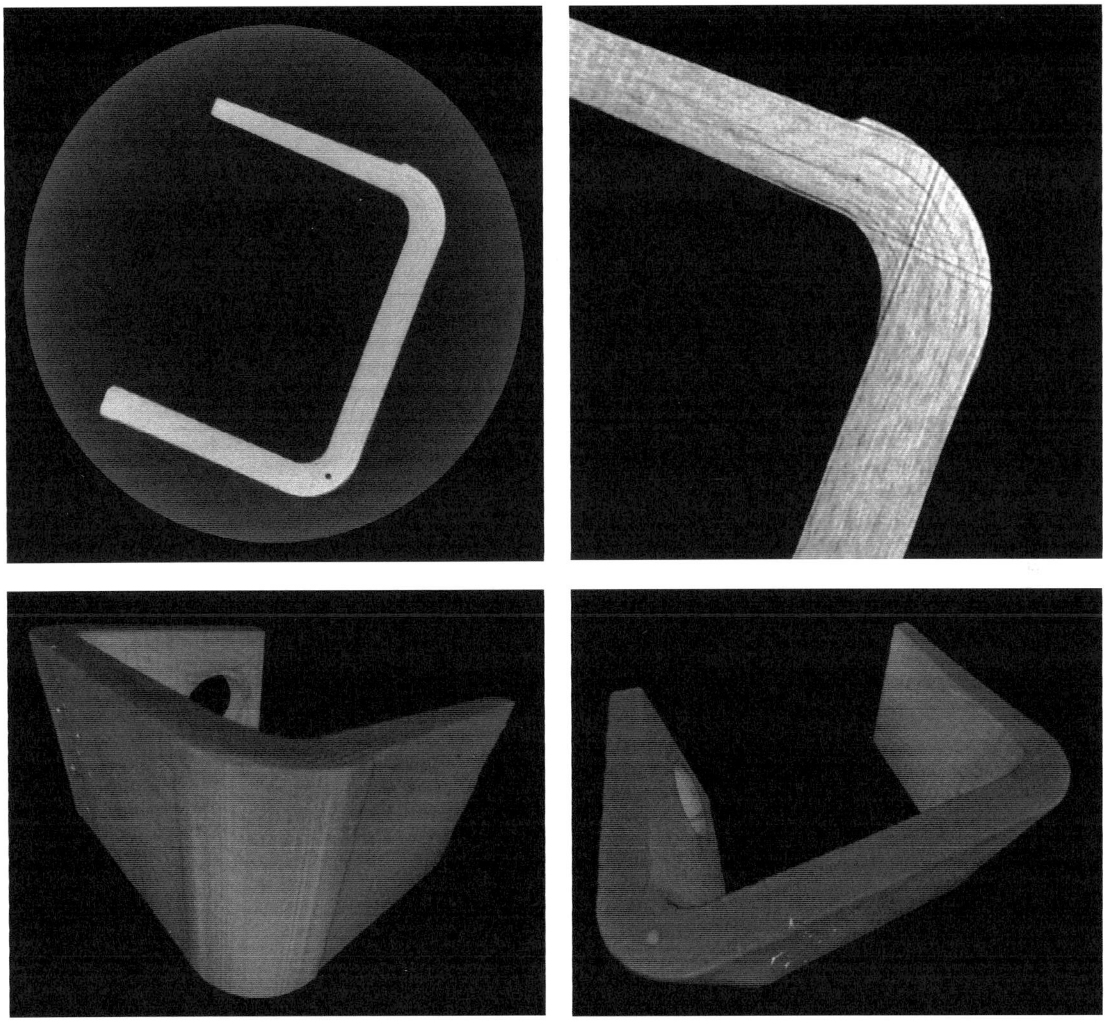

Figure 3. Three-dimensional models of the beam segment of the U-shaped profile.

The design and implementation of the system being developed will allow low- or multi-angle survey of inspected objects, which will significantly increase the reliability of the detection of inhomogeneities.

IOP Conf. Series: Journal of Physics: Conf. Series **967** (2018) 012007 doi:10.1088/1742-6596/967/1/012007

Acknowledgments
The research was carried out at the Perm National Research Polytechnic University with the support of the Russian Science Foundation (project No. 16-19-00155).

References

[1] Potrakhov N N, Zhamova K K, Bessonov V B, Gryaznov A Y and Obogovsky A V 2015 *PNRPU Aerospace Engineering Bulletin* **43**
[2] Gagauz F M 2012 *The design and production of aircraft constructions. Collection of scientific papers KhAI* **4** 15–20
[3] Anoshkin A N, Osokin V M, Tretyakov A A, Potrakhov N N and Bessonov V B 2017 *Journal of Physics: Conference Series* **808** 012003
[4] Rumyantsev S V 1974 *X-ray flaw detection* (Moscow: Atomizdat)

Circuit modification and research of operation modes of high-frequency pulsed resonant converter of the X-ray tube power supply

V V Klonov, I A Larionov and V B Bessonov

Saint-Petersburg Electrotechnical University "LETI", 197376, Saint-Petersburg, Russia

E-mail: vbbessonov@yandex.ru

Abstract. Despite obvious drawbacks of the resonant converter, such as complicated calculation, increased size and weight of the device, deviations of the circuit parameters from product to product, the resonant converter shows significant advantages in comparison with other. The task was to design the generator, which is built on a resonant topology.

In the modern market of switching power supplies resonant converters find a narrow scope. This is due to the fact that the process of the development of classical PWM converters is relatively well-established and researched. Therefore, apart from the fact that existing knowledge of the processes in the converter and the years of experience of their calculation and design significantly simplify and accelerate the development of modern power source, circuits, based on the PWM control, are simple and therefore more reliable.

However, despite obvious drawbacks of the resonant converter, such as complicated calculation, increased size and weight of the device, deviations of the circuit parameters from product to product (forcing to carry out a manual configuration of each individual power source), the resonant converter shows significant advantages in comparison with other. These include: lower dynamic losses in power switches, reduced electromagnetic interference, increased efficiency in the use of the transformer core and output rectifier or multiplier [1].

The task was to design the generator, which is built on a resonant topology. The first result was a layout, witch tested on a low output voltage (transformation ratio for simplicity, was taken 1). The Converter was based on LC topologies and showed a good result at the load power 320 W, temperature of switches and reactive elements did not rise above 55 °C. A block diagram of the control shown in figure 1, demonstrates the principle of source control. A microcontroller was selected as generator of control signal because it has greatly simplified the operation mode control and debug from a personal computer, but also increased system reliability, simplified the circuit and improved the ability of control the operating mode of the transmitter: adjustment in a wide range of frequency and duty cycle of the bridge circuit to create the possibility of algorithms of automatic tuning of parameters and calibration. The disadvantage of the use of microprocessor-based systems include the inability to stabilize the output voltage with fast transient response. In addition, the microcontroller is a digital system, so the frequency and pulse width adjustment is limited by the discretization of the internal timers of the chip.

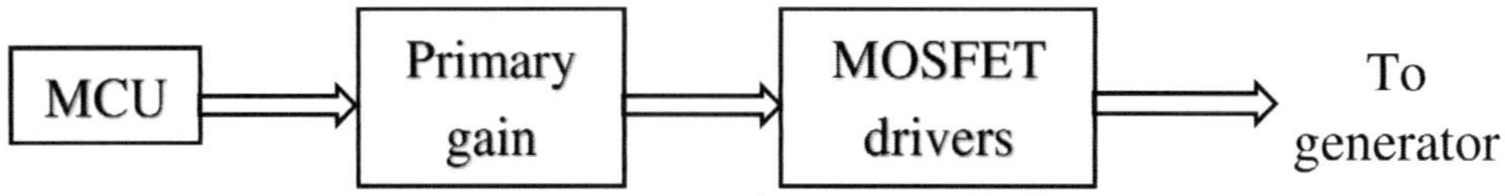

Figure 1. Block diagram of the converter.

As the input source uses a stable source of 24 V. The adjustment was carried out by a frequency method. With increasing frequency LC filter, which has certain resonant frequency, selects all a smaller part of the main harmonic of the primary signal. The voltage drop at switches at the turn-on time is equal to the voltage drop across the diode that eliminates switching losses. The opposite effect is switching at zero current, when the load is capacitive (frequency below the resonance) and transistors turn-on is lossless. The second method has several disadvantages (increased current load on the diodes, the presence turn-on losses because of transistor's internal capacitance), so the resonant converters typically operate at frequencies above the resonance frequency of the LC circuit. The exception is topologies, built on the current devices, such as IGBT, which has a "current tail", or the thyristors, which turn-off will happen in a natural way and without losses (the current through the switch itself reaches zero).

When testing circuits on high-voltage transformer (schematic figure 2) the behavior of the scheme has changed its character. First, the spectral composition of the output voltage in the primary coil was significantly reduced in all investigated frequency range (60–120 kHz, the true resonant frequency of the transducer 72 kHz) the output signal was an almost pure sine. Secondly, it changed the load effect on the resonant frequency of the circuit: on low-load resonant frequency is higher than high that were opposite to the results obtained previously. Thirdly, PWM adjustment (the test was conducted in the range of duty cycle 30–95 %) are now not only distorts the output signal (distortion starts only in the region of small fill factor), but almost linearly adjusts the amplitude of the voltage on the primary and, as a consequence, the secondary winding.

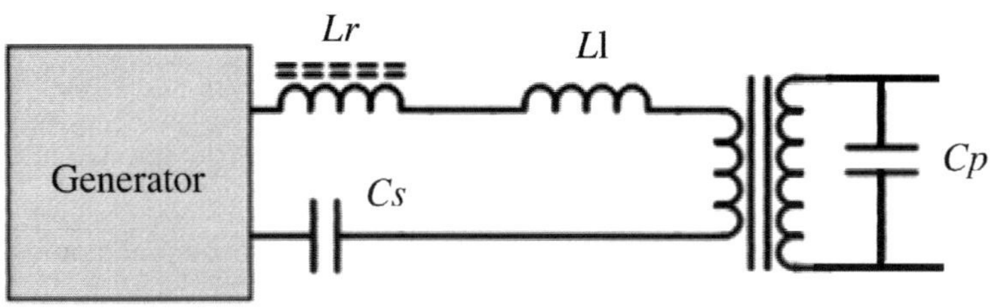

Figure 2. Electrical circuit of resonant converter.

In the case of high voltage transformer, the secondary winding, because of their large number, form a sufficiently large parasitic capacitance (evaluation value varies in the region of tens of nF). This capacity is connected in parallel to the primary winding. On the side of the primary circuit, the capacity is increased in accordance with the transformation ratio. The obtained scheme is another rather famous variant of resonant topology and is called the series-parallel resonant LCC Converter.

A distinctive feature of the LCC circuit is the increase of the resonance frequency with decreasing load. This is due to the fact that in a resonant circuit contributes to the parallel capacitance, reducing the effective capacity of the LC circuit in accordance with the formula (1):

$$Cr=\frac{Cp\cdot Cs}{Cp+Cs}, \qquad (1)$$

where Cr – resulting resonant capacitance, F; Cs – serial capacitance, F; Cp – parallel capacitance, F.
Distinctive features of the LCC Converter are:

– the possibility of voltage gains. At that time, as LC converter operates as a buck converter, a LCC circuit allows to obtain a voltage gain less than or greater than one;

IOP Conf. Series: Journal of Physics: Conf. Series **967** (2018) 012008 doi:10.1088/1742-6596/967/1/012008

– adjustment of *Cp* allows to adjust gain and q-factor of resonant LCC circuit: low q-factor allows to simplify the frequency adjustment by reducing the steepness of the gain curve;

– LCC Converter provides a cleaner spectral composition of signal on the windings of the transformer, which allows using cores with lower cut-off frequency (and thus greater magnetic permeability);

– in parallel and series-parallel topologies, there is a current flowing through a parallel reactivity and increase the consumption of the scheme: on the one hand, this allows to obtain voltage gain, with another – increases load current [2].

Modification of the power generator to the resonant LCC Converter was carried out. Figure 3 shows a photo of the power board that generates an alternating voltage across the primary winding of the transformer. Since the peak current load on the inductance is about 24 A, it is necessary to take special measures to protect the component from saturation. Magnetic induction in the annular core can be simplistically calculated by the formula (2):

$$B = \frac{L \cdot I}{N \cdot S} = \frac{Al \cdot N \cdot I}{S}, \tag{2}$$

where B – magnetic induction value in the core, T; I – current through the coil, A; L – inductance of the coil, H; N – number of turns, Al – specific inductance of the core, H/turn^2; S – effective cross-sectional area of the core. Thus, to reduce the load on the component (in addition to the protection of saturation the decrease of B leads to lower losses in the core) reduced the number of primary turns of the transformer. To compensate the additional inductance and reduce a magnetic field in the coil cross-sectional area was increased by combining the three ring cores. In addition, film capacitors of 22 nF was added to reduce dynamic losses when switches turn-off, and external Schottky diodes was added, having the best time and static characteristic than the body diodes of MOSFET transistors. Since the PWM adjustment reduces the load on the transistors and increase the diodes load, diodes in TO-220 package and with a current of 20 A are used in application.

Figure 3. Power board of resonant converter and a modified design of high voltage transformer.

A negative consequence of resonant inductance reducing is increased reactive current, especially at high frequencies. This reduces the efficiency of the circuit, therefore the main objective for optimization of efficiency of converter is parallel capacity reduction, i.e. self-capacitance of the transformer's secondary windings. It is worth noting that in classical PWM generator this capacity has a more destructive effect – it leads to inrush current in the switching transistors and, consequently, to serious dynamic losses. The only method of reducing the winding capacitance of the transformer is modification of its design. The main directions in reducing of winding capacitance includes:

– separation of turns with a large potential difference;

– increasing the thickness of the wire insulation;

– separation of the primary and secondary windings.

Also figure 3 shows a photo of a modified design of high voltage transformer. The basically was taken design with partitioned windings: the winding area is divided into sections [3]. This approach provides a uniform increasing of potential from one end of coil to the other and, as a result, reduced capacity. The coil is made of PTFE. After winding the secondary winding (it is wrapped around the first due to minimal parasitic inductance on secondary side) the PTFE case is put on and the primary winding is wound. Formed due to the large air gap parasitic inductance is a part of resonant inductance and does not contribute significantly to the resonance frequency due to the small number of turns (practical scheme on the primary side used 2–3 turns).

The resonant converter was tested on high-voltage transformer, obtained satisfactory results, associated with the increasing of the circuit efficiency. It was increased due to the following factors:

– reduced switching losses due to the additional capacitors on the keys;
– body diodes of transistor were replaced by external, more efficient Schottky diodes;
– design of the transformer and the resonant inductance were optimized.

Acknowledgments

Work is performed with financial support and at the expense of a grant of the President of the Russian Federation for the state support of young Russian scientists MK-1582.2017.8.

References

[1] Kolpakov A 2000 *In the labyrinth of power electronics*
[2] *Philips semiconductors. Power semiconductor applications – Switch mode power supplies.* URL: http://educypedia.karadimov.info/library/APPCHP2.pdf.
[3] Vdovin C 1971 *Design of pulse transformers*

IOP Conf. Series: Journal of Physics: Conf. Series **967** (2018) 012009 doi:10.1088/1742-6596/967/1/012009

Microfocus computed tomography in medicine

A V Obodovskiy

Saint Petersburg Electrotechnical University "LETI", 197022, Saint-Petersburg, Russia

E-mail: obodovsky@yandex.ru

Abstract. Recent advances in the field of high-frequency power schemes for X-ray devices allow the creation of high-resolution instruments. At the department of electronic devices and Equipment of the St. Petersburg State Electrotechnical University, a model of a microfocus computer tomograph was developed. Used equipment allows to receive projection data with an increase up to 100 times. A distinctive feature of the device is the possibility of implementing various schemes for obtaining projection data.

The development of technology has led to a significant increase in the computing power of electronic devices. Modern achievements open new opportunities in the design of devices for X-ray diagnostics. The development of X-ray diagnostics in medicine deserves special attention. X-ray tomography is the most actively developing area.

Recent advances in the field of high-frequency power schemes for X-ray devices allow the creation of high-resolution instruments. The cone-beam scheme is most preferable when creating similar systems. An illustration of such a scheme is shown in figure 1. New dynamic X-ray detectors have dimensions up to 400×400 mm. Such dimensions allow tomography studies of individual parts of the human body, for example, the head. Such receivers can receive images with a frequency of up to 30 frames per second.

The natural limitation is the blurriness of the resulting X-ray images [1]. This is due to the dimensions of the effective focal spot of the X-ray tube. The dimensions of the focal spot range from 0.5 to 1.5 mm in classical X-ray machines. Individual parts of the patient's body will be blurred during research, which leads to the formation of artifacts. This disadvantage can be eliminated by using a microfocus X-ray machine [2]. This will make it possible to use X-ray optical schemes with direct geometric magnification. Obtaining tomograms with a voxel size of less than 50 μm becomes possible under special shooting conditions.

At the department of electronic devices and Equipment of the St. Petersburg State Electrotechnical University, a model of a microfocus computer tomograph was developed. Used equipment allows to receive projection data with an increase up to 100 times. A distinctive feature of the device is the possibility of implementing various schemes for obtaining projection data.

A series of experiments allowed to determine the minimum necessary amount of projection data for a qualitative restoration of tomograms. This value was not less than 200 projections in the range of angles from 180 degrees. The capabilities of modern X-ray receivers allow images to be obtained at a high frequency. This makes it possible to obtain the necessary data in 10 seconds or less. In this case, the patient is under the influence of radiation less than when receiving orthopantomograms.

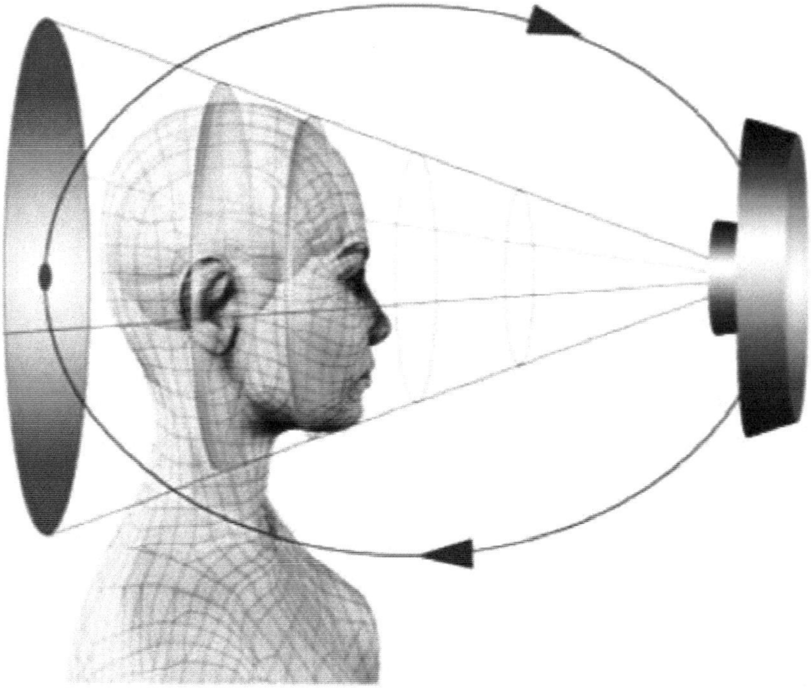

Figure 1. Cone-beam scheme for obtaining tomograms.

The internal structure of the object of investigation is restored by the method of reverse projection with filtration. The processing speed of the projection data is very high due to the use of calculations on a high-performance graphics processor. The speed increase is up to 50 times when using parallel calculations.

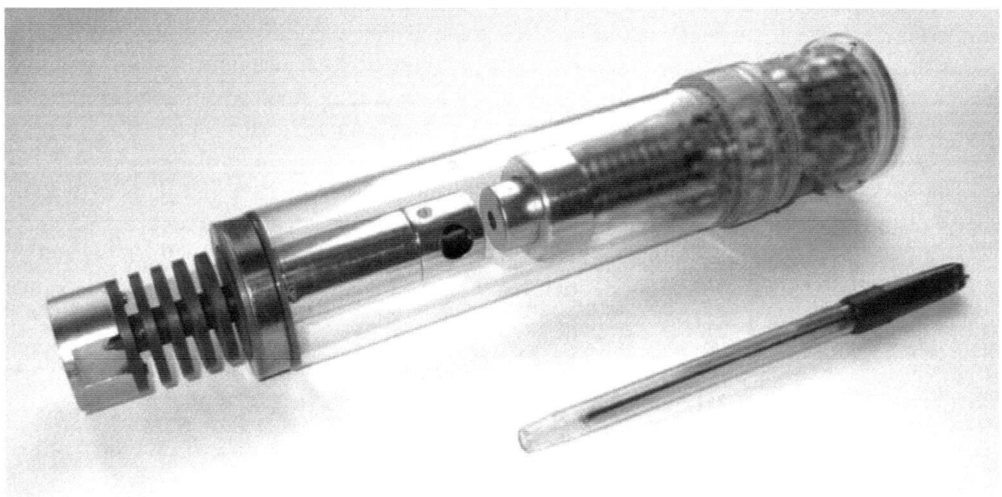

Figure 2. Appearance of the X-ray tube with photocathode.

The next step is to create a special X-ray source based on a tube with a photocathode to improve the equipment for performing tomographic studies [3]. The appearance of such an X-ray tube is shown in figure 2. Such a radiation source will reduce the time of exposure to radiation. This will significantly reduce the radiation burden on the patient.

IOP Conf. Series: Journal of Physics: Conf. Series **967** (2018) 012009 doi:10.1088/1742-6596/967/1/012009

References

[1] Blinov N N, Vasilyev A Y, Gryaznov A Y, Zhamova K K, Bessonov V B, Potrakhov E N and Potrakhov N N 2014 *Medical equipment* **2** 2–4

[2] Potrakhov N N, Gryaznov A Y, Zhamova K K, Bessonov V B, Obodovskiy A V, Staroverov N E and Holopova E D 2015 *Biotekhnosfera* **5** 55–63

[3] Bessonov V B, Gryaznov A Y, Zhamova K K, Potrakhov E N, Potrakhov N N and Shishov D I 2014 *Mashinostroitel* **12** 46–9

Shift focal spot X-ray tube to the imposition anode under long exposure

A V Obodovskiy, V B Bessonov and I A Larionov

Saint Petersburg Electrotechnical University "LETI", 197022, Saint-Petersburg, Russia

E-mail: obodovsky@yandex.ru

Abstract. X-ray non-destructive testing is an integral part of any modern industrial production. Microfocus X-ray sources make it possible to obtain projected images with an increased spatial resolution by using a direct geometric magnification during the survey. On the basis of the St. Petersburg State Electrotechnical University staff of the department of electronic devices and equipment has been designed model of microfocus X-ray computed tomography.

X-ray non-destructive testing is an integral part of any modern industrial production. One of the main trends in the development of X-ray defectoscopic complexes is the improvement of spatial resolution capabilities. Microfocus X-ray sources make it possible to obtain projected images with an increased spatial resolution by using a direct geometric magnification during the survey [1].

This distinctive feature of microfocus X-ray machines can be used to obtain tomographic images. This gives rise to the development of a separate field of technology – microfocus X-ray tomography or microtomography [2].

The large value of the magnification factor in the microfocus tomography imposes serious demands on the accuracy of positioning of all elements of the tomographic system. The main factors affecting the quality of tomographic image restoration:

1. The angle of inclination of the rotation axis of the object to the vertical axis of the radiation receiver;
2. Accuracy of measuring the distance from the source to the X-ray receiver;
3. Shift of the X-ray source coordinate relative to the line connecting the axis of rotation and the geometric center of the radiation receiver.

The low accuracy of positioning the listed elements leads to the formation of various kinds of "artifacts" on the image being reconstructed and can cause distortion of small parts and defects of the objects under study.

On the basis of the St. Petersburg State Electrotechnical University staff of the department of electronic devices and equipment has been designed model of microfocus X-ray computed tomography (figure 1). Monoblock microfocus apparatus based on the sealed X-ray tube BS-16 was used as an X-ray source (figure 2). The main technical characteristics of the tube used are shown in table 1.

The small size of the focal spot is achieved by using a double focusing system of the electron beam. This makes it possible to obtain X-ray images with geometric magnification up to 200 times. The effective size of the voxel of tomographic images reaches 1 μm.

IOP Conf. Series: Journal of Physics: Conf. Series **967** (2018) 012010 doi:10.1088/1742-6596/967/1/012010

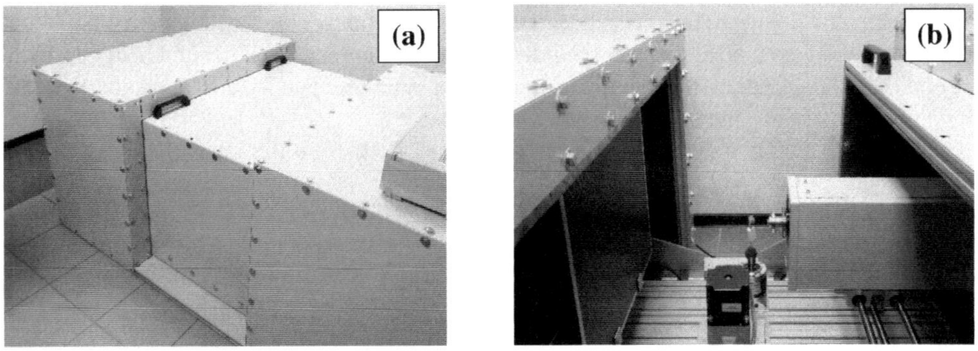

Figure 1. Microfocus X-ray computer tomograph MRKT-01: **(a)** – appearance; **(b)** – source and receiver of radiation with a displacement system.

Table 1. Technical specifications.

Anode voltage, kV	40–150
Anode current, mA	0.01–0.2
Maximum power in continuous mode, W	7
Focal spot size, μm	<5
Mode of operation	continuous, re-short-term
Anode performance type	impaired
Dimensions (L×W×H), mm	315×76×76
Current of heat, A	3.9

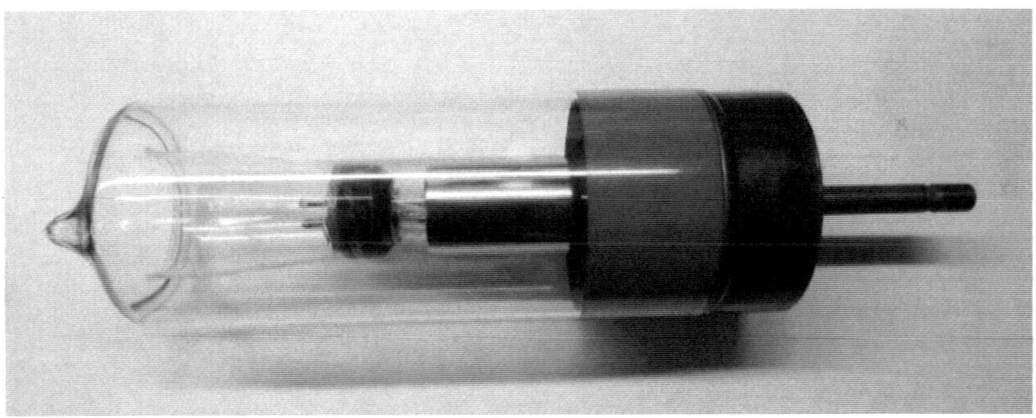

Figure 2. X-ray tube of the closed type BS-16.

The X-ray apparatus works continuously for a long time. For example, it takes more than 26 min of continuous operation for a set of 800 projections and an exposure time of 2 s for each projection. Various specific effects begin to appear in similar modes of operation of the X-ray tube. One such effect is the displacement of the focal spot over the surface of anode. Such a shift does not have a significant effect on the resulting images in single images at small magnification ratios. The effect becomes significant when typing projection data for a tomography with a large magnification. This leads to a dynamic disruption of the positioning accuracy of the elements of the tomographic system.

The endings of two metal needles 0.7 mm in diameter were taken to study the characteristic dimensions of the focal spot displacement. The needles were located close to each other. The direct geometric increase was 200 times (figure 3). The survey was carried out at an accelerating voltage of 115 kV and an anode current of 20 μA. In all, 200 shots were taken with an exposure of 1 s for each shot. The coordinates of the point of each needle were determined on the obtained images and the deviation from the original position was recorded. The calculation of the deviation of the focal spot was made taking into account the magnification factor and the pixel dimensions of the radiation receiver. The results of the calculations are shown in figure 4.

The results of the studies indicate the presence of a direct pattern of the magnitude and direction of the focal spot drift. Two ways to compensate for the drift of the focal spot are proposed based on the experiments performed:

1. Use of an electromagnetic deflection system mounted on the anode of an X-ray tube;
2. Digital displacement of the image after its registration, taking into account the position of the reference points in the original picture.

Further research will continue in the direction of a quantitative analysis of the effect of the effect examined on the reconstruction of a tomographic image.

Figure 3. X-ray image of two needles.

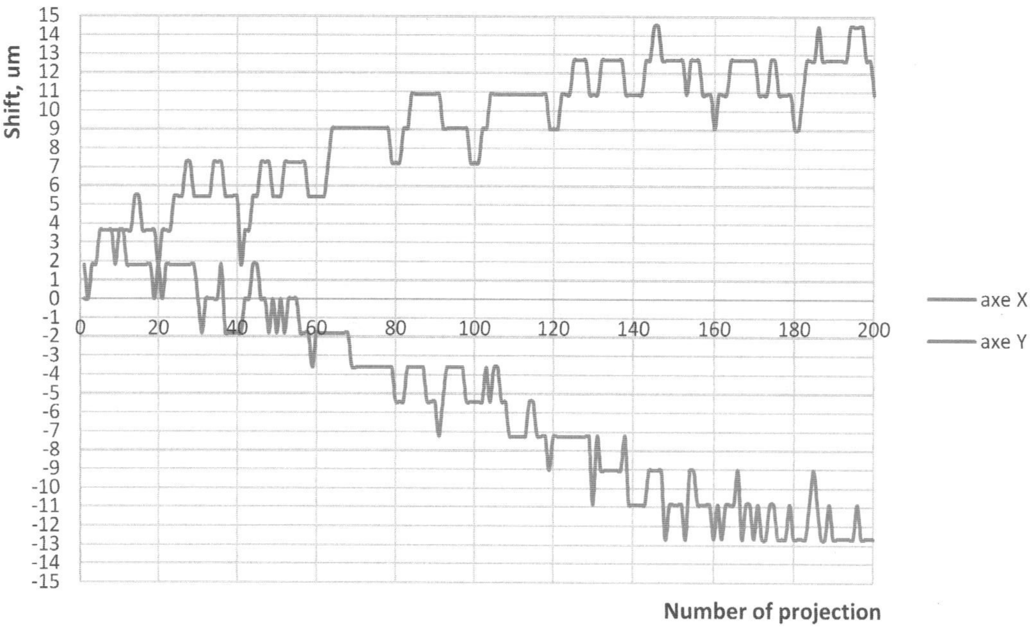

Figure 4. Dependence of the amount of displacement of the focal spot on the projection number.

Acknowledgments

The work was supported by the Russian Science Foundation in the framework of the project "Creation of a portable installation for microfocus radiography for the purpose of operative control of microstructure, physical and chemical properties and determination of the remaining life of aviation components and assemblies of polymer composite materials". Project number is 15-19-00259.

References

[1] Bessonov V B, Obodovskiy A V, Klonov V V and Kostrin D K 2014 *The Eurasian Union of Scientists* **5/3** 12–5

[2] Potrakhov N N, Gryaznov A Y, Zhamova K K, Bessonov V B, Obodovskiy A V, Staroverov N E and Holopova E D 2015 *Biotekhnosfera* **5** 55–63

Protection of the electronic components of measuring equipment from the X-ray radiation

N O Perez Vasquez, D K Kostrin and A A Uhov

Department of electronic instruments and devices, Saint Petersburg Electrotechnical University "LETI", 197376, Saint Petersburg, Russia

E-mail: dkkostrin@mail.ru

Abstract. In this work the effect of X-ray radiation on the operation of integrated circuits of the measurement equipment is discussed. The results of the calculations of a shielding system, allowing using integrated circuits with a high degree of integration in the vicinity of the X-ray source, are shown. The results of the verification of two measurement devices that was used for more than five years in the facility for training and testing of X-ray tubes are presented.

When X-ray or γ-ray radiation passes through a substance part of their energy is absorbed by the material, resulting in ionization of atoms of this substance. Thus X-ray radiation is one of the types of the ionizing radiation [1]. The ionizing radiation leads to changes in properties of matter due to the violations of their structure (breaking of chemical bonds, formation of free radicals, coloration, etc.).

The dominant mechanism of action of the pulsed ionizing radiation on the semiconductor devices and integrated circuits (ICs) is the volumetric ionization of semiconductor regions, i.e. the generation in them of nonequilibrium electron–hole pairs. Diffusion and drift of the generated by the radiation nonequilibrium charge carriers results in the temporary or permanent changes in the device parameters, which may lead to the device malfunction [2]. The ionization reaction is defined by the primary and secondary volumetric ionization effects that occur under the exposure to pulsed radiation.

Primary volumetric ionization effects occur due to the ionization of the different areas of IC and the associated short-term (commensurate in duration with working times) changes in the parameters of the devices. Secondary ionization effects are determined by reaction of the irradiated object to the ionization (increase of the ionization current, "latch-up" effect, induced secondary breakdown, etc.) and can have a long-term character.

The dominant mechanisms of failure in the CMOS ICs are connected with ionization effects. Surface ionization effects are mainly related to the accumulation of charges in thin layers of the passivating dielectric under the gate electrode and at the interface of Si/SiO_2. These effects are dominant when exposed to a stationary ionizing radiation (gamma, X-ray and electronic).

Of particular interest is the impact of ionizing radiation on operational amplifiers (op-amps), because they are the electronic components necessary to create measuring devices. Impact of the ionization effects is primarily in the form of a significant increase in the leakage and channel currents, which leads to an increase of input offset and shift currents.

Specialized operational amplifiers, including circuits with high input impedance, precision micropower op-amps and high-speed amplifiers, are usually more sensitive to residual radiation effects, as circuit design and technological measures used to achieve extreme performance for any

IOP Conf. Series: Journal of Physics: Conf. Series **967** (2018) 012011 doi:10.1088/1742-6596/967/1/012011

parameters usually reduce their radiation resistance. The best option to work under the influence of the ionizing radiation is the elements with low and medium degree of integration, based on analog or combinational elements. If there is a need to use microntroller (μC) and other complex ICs (such as analog-to-digital converter or ADC) it is required to take additional precautions, since information stored in their memory is under a threat of removal.

Current converter must operate in conditions where, depending on the type of the X-ray tube, the maximum quantum energy does not exceed 320 keV [3]. Device protection can be a screen of iron and lead, which is simultaneously the housing of the device. The casing is a metal shell made of stainless steel with a wall thickness of 0.5 mm, in which the batteries and lead boxing, where a PCB with ICs is mounted, are placed (figure 1). The box is made of lead plates with a thickness of about 5–6 mm.

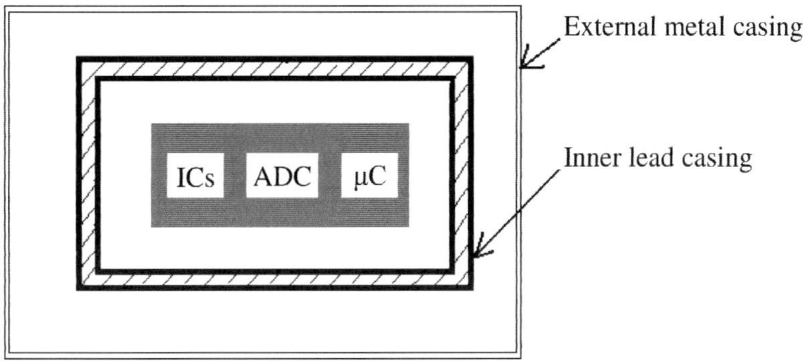

Figure 1. Protective housing of the current converter.

The calculation showing the degree of attenuation of radiation was made for the X-ray tube with anode made of tungsten for the maximum operating current of 50 mA. Calculations were made on the basis of developed in SPbETU "LETI" techniques [4] with the help of Mathcad software (figures 2, 3). It was assumed that X-rays must pass through: beryllium window, air, iron filter and lead filter before irradiating the ICs, μC and ADC.

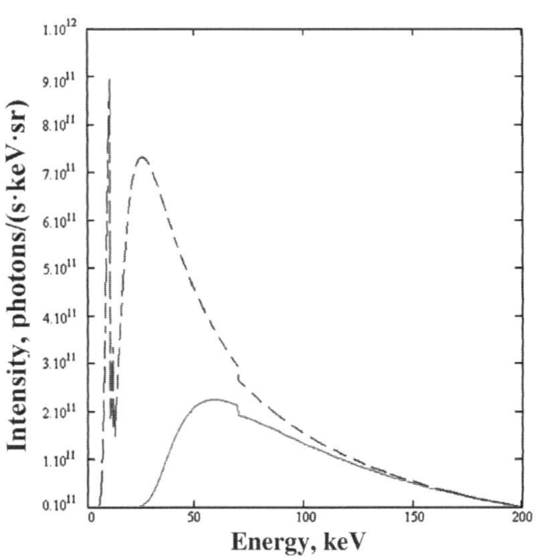

Figure 2. Spectra of a braking X-ray radiation: before (dashed line) and after (solid line) the steel filter.

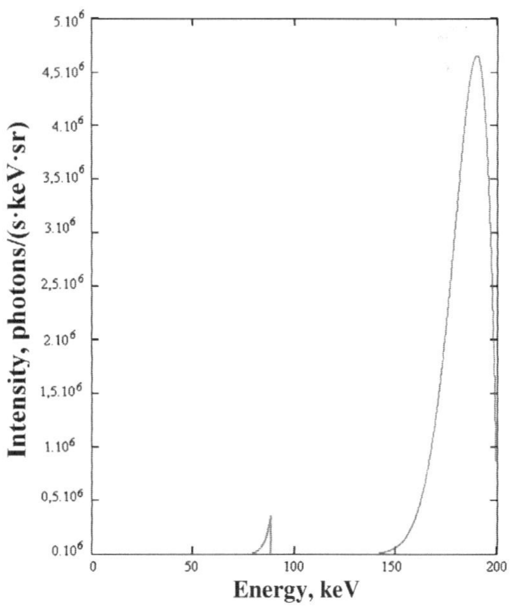

Figure 3. Spectrum of a braking X-ray radiation after the lead filter with thickness of 5 mm.

The calculation was carried out using expressions describing the spectrum of a braking X-ray radiation according to the formula of Kramers [5] taking into account attenuation in: target of the X-ray tube, exhaust (beryllium) window, two filters, installed on the beam path in the air gap between the tube and the detector. Graphs of the received functions are displayed in figures 2, 3.

From calculations it can be seen that the layer of steel attenuates X-ray radiation by about 2.5 times, but only the radiation with quantum energy less than 50 keV is attenuated. Hard radiation is almost completely not absorbed in the metal layer.

Figure 3 shows spectrum of a braking X-ray radiation passed through the two screens – made of steel and lead. The graph shows that ionizing radiation is almost completely absorbed in a lead filter with a thickness of 5 mm. Calculation shows that the radiation passing through the lead filter is weakened by $1.636 \cdot 10^5$ times, and the total absorption coefficient of two filters is $4.107 \cdot 10^5$.

Thus, this X-ray equipment can cause damage to the conversion module, so the measures of protection against the ionizing radiation confirm its necessity. The harm that may cause the ionizing radiation is the increase in the generation currents, which can lead to undesirable errors, growth of the space charge in thick SiO_2 layers, which can lead to increased power consumption, and various kinds of spurious connections between the internal cells of integrated circuits.

During the calculation was made assumption that the metal plates (filters) are directly placed in the path of the X-ray beam. In fact, they are not, and the microammeter stands apart from the main beam of radiation, so the device gets even fewer quanta of radiation and, accordingly, there will be less absorbed radiation (figure 4).

Figure 4. Placement of the measuring devices and X-ray tubes in the test chamber.

When training X-ray tubes for the medical diagnosis of BD-150 series with a rotating anode in the CJSC "Svetlana-X-Ray" is used a pulsed high voltage power supply, assembled on a three-phase bridge rectification circuit [6]. On the trained X-ray tube voltage pulses with amplitude from 10 to 250 kV are provided with the duration from 10 ms to 10 s. Under these conditions, the amplitude of the X-ray tube current can vary from 5 to 500 mA and the potential at which the measuring circuit is located can be up to 250 kV.

Analysis of the factors determining the total error of the developed measuring devices under the stable external conditions showed that theoretically the relative error of measurement may be below 1 %. After five years of operation of these devices under the influence of the high voltage electric field and X-ray radiation a test was conducted to check the amplitude transmission characteristics of the developed ammeters (figure 5).

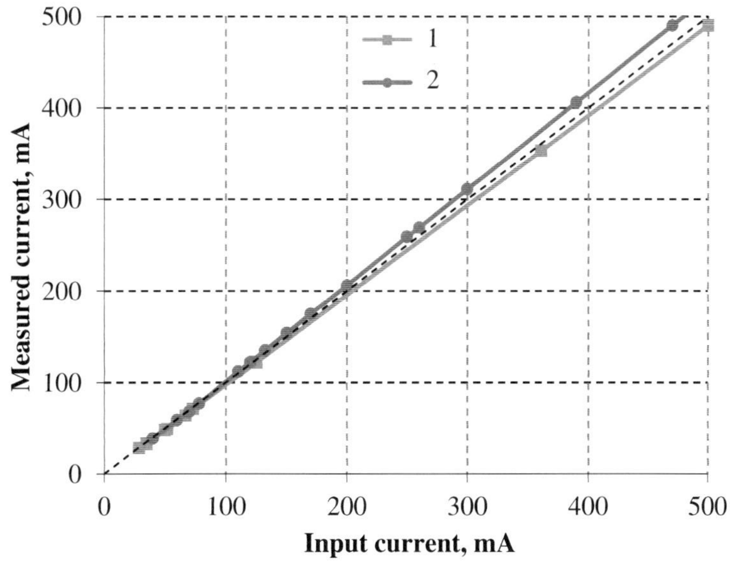

Figure 5. Test results for two measuring devices after 5 years of operation.

As the result, it was found that the measurement error after 5 years of operation have deteriorated to 5 % at the end of the conversion scale, that is still acceptable during training of the X-ray tubes. As can be seen from figure 5 response of the instrument retained after 5 years of operation its linearity and only a small change in the angle of inclination of the characteristics can be seen, which can be easily fixed by adjusting the calibration constants.

References

[1] Shirshev L G 1969 *Ionizing radiation and electronics* (Moscow: Sovetskoe radio)
[2] Vavilov V S and Uhin N A 1969 *Radiation effects in semiconductors and semiconductor devices* (Moscow: Atomizdat)
[3] Perez Vasquez N O, Kostrin D K and Uhov A A 2017 *Journal of Physics: Conference Series* **808** 012007
[4] Lukianchenko E M and Gryaznov A Yu 2003 *Proceedings of SPb ETU "LETI"* **8** 10–4
[5] Kliuev V V 1993 *X-ray technique* (Moscow: Nauka)
[6] Bystrov Yu A, Kostrin D K, Peres Waskes N O and Ukhov A A 2013 *Biotechnosphere* **4** 60–4

X-ray examinations of newborns

N N Potrakhov and Y N Potrakhov

Saint Petersburg Electrotechnical University "LETI"

E-mail: yn@eltech-med.com

Abstract. At the present time, the basis of instrumental diagnostics of atelectasis is lung radiography. In the case of preterm infants, it should be performed immediately after birth, and then regularly throughout the entire nursing period. The purpose of the project, within the framework of which this development is carried out, is the creation of an original domestic digital low-dose technology for X-ray examinations in neonatology, including in non-stationary conditions.

Modern medical practice shows that up to 10 % of the total number of newborns are born annually prematurely because of spontaneous premature births or because of planned artificial termination of pregnancy in later periods.

According to the definition of the World Health Organization since 1974, and in Russia since 2013 [Order of the Ministry of Health and Social Development of Russia No. 1687n], newborns are considered viable at a body weight of 500 g and above. Such a child fits in the palm of an adult and, unfortunately, at the time of birth has a whole set of diseases. One of the most dangerous is atelectasis of the lung – incomplete expansion of the lung tissue in a restricted area or in the entire lung.

At the present time, the basis of instrumental diagnostics of atelectasis is lung radiography. In the case of preterm infants, it should be performed immediately after birth, and then regularly throughout the entire nursing period. At the same time, during nursing, the child is in a special box – an incubator (cuvette), in which the necessary regime of temperature, humidity, oxygenation, feeding and introducing the necessary drugs through special probes is maintained. Accordingly, such children should not be X-rayed in stationary X-ray rooms, but in maternity halls, emergency units, intensive care units and, in some cases, during transportation of the newborn from the maternity home to the perinatal center.

The radiographic methods standardized in Russian medicine not allows conducting X-ray studies of premature babies in the described – unspecialized conditions. The main reasons for this are:

– the lack of technical diagnostic tools that make it possible to obtain X-ray images of such specific objects as the thoracic area of a child weighing 500 g, having the necessary informativeness;

– the lack of methods for conducting X-ray studies in neonatology in unspecialized conditions, including methods of patient placement, mode selection tables, etc.

It should be specially noted, that at present time there are no commercially available domestic or foreign X-ray machines that allow conducting X-ray studies in the described conditions.

At the same time, conducting digital radiography for newborns is very much in demand. According to the Children's City Hospital No.1 (St. Petersburg), only in its intensive care units about 4500 roentgenological procedures are performed every year in 2 or 3 projections. In total, up to 10 thousand X-ray images per year are performed.

Currently, these studies are conducted on domestic or foreign X-ray machines that are practically not adapted to the tasks and conditions of resuscitation of newborns with pathologies. Accordingly, the medical personnel and the nearby patients are subjected to an additional radiation load.

The purpose of the project, within the framework of which this development is carried out, is the creation of an original domestic digital low-dose technology for X-ray examinations in neonatology, including in non-stationary conditions. Studies conducted in conjunction with specialists from the Almazov National Medical Research Center of the Ministry of Health of the Russian Federation and of the Children's City Hospital No.1 showed that practically all the listed requirements for organizing X-ray diagnostic studies in unspecialized conditions and, first of all, neonatology, are met by Microfocus X-ray radiography [1]. Its main distinguishing features are:

– the use of X-ray machines with a focal spot size of an X-ray tube of less than 0.1 mm;

– 1.5–2 fold increase in voltage at the X-ray tube and a decrease in the focal distance of 3–5 times as compared to standard X-ray technique [2, 3].

To implement this method in the first stage of the project, a specialized digital X-ray diagnostic complex was used as a prototype, including the X-ray device of the family PARDUS and X-ray imaging systems based on a photo-stimulated screen in the portable design. The design of the first domestic portable device "PARDUS-R" allows you to obtain X-ray images without the use of a stationary tripod. The X-ray staff performs the survey, holding the device in its hand and being in a radiation-safe environment (figure 1) [4].

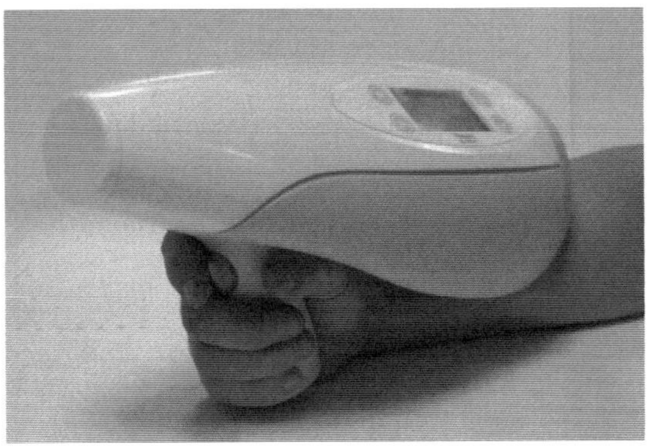

Figure 1. Portable X-ray device "PARDUS-R".

The approbation of the technique was carried out on the basis of a multidisciplinary children's hospital. Figure 2 shows X-ray images of the thoracic region (Patient P., age – 7 months, weight – 3.5 kg) obtained in a stationary X-ray cabinet (figure 2(a)), and obtained "at home" in the same patient (figure 2(b)) using the prototype of the complex.

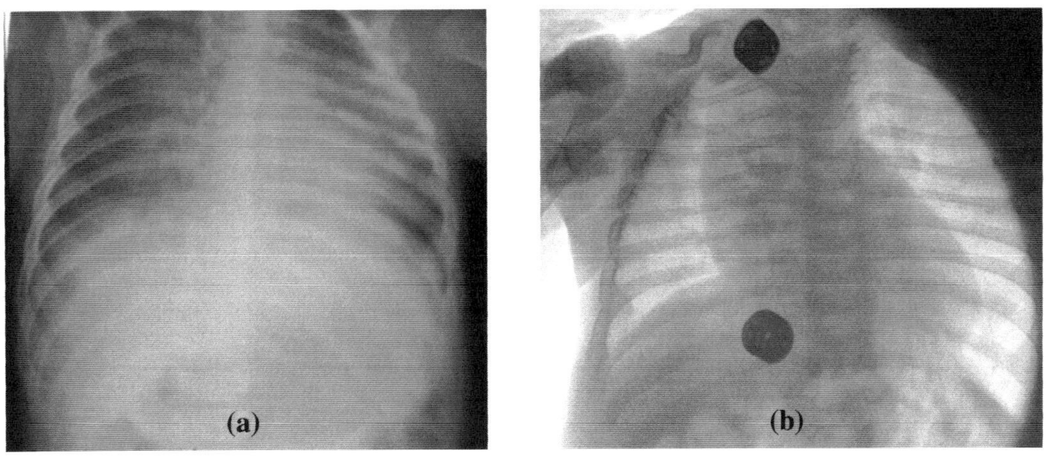

Figure 2. X-ray images of the chest of the newborn.

On X-ray images obtained under non-stationary conditions, the lung tissue, the contours of the heart and the dome of the diaphragm are clearly traced. Figure 3 shows a picture of the lumber region, also obtained in non-stationary conditions (in the ward).

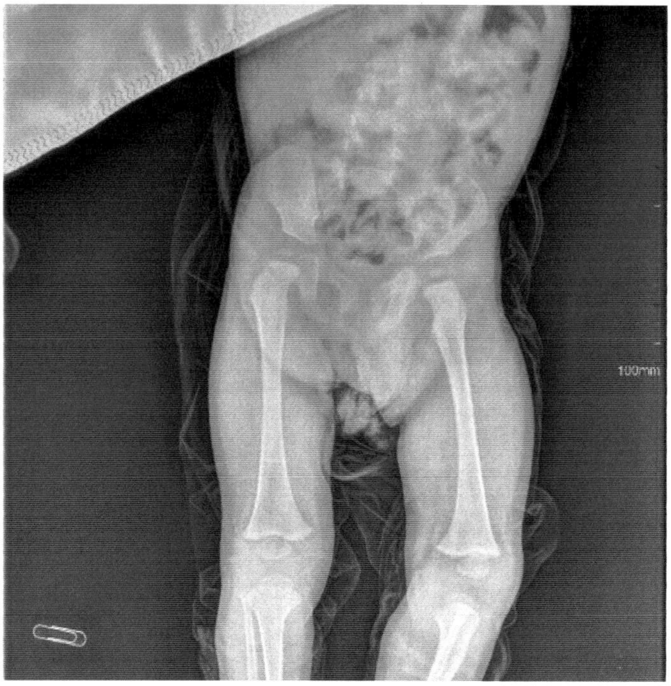

Figure 3. X-ray image of the newborn's lumber.

All bone structures are visualized clearly. According to experts, the information content of all X-ray images obtained in non-specialized and non-stationary conditions is almost not inferior to the X-ray image obtained in the hospital.

Thus, the results of the studies open up the prospect of using the technique of Microfocus X-ray radiography with portable technical means for researching newborns.

Acknowledgments
The work is supported by the Federal Targeted Program "Research and Development in Priority Areas for the progress of the Russian Science and Technology Complex for 2014-2020" (Event 1.3) under Contract No. 14.578.21.0262 "Digital acute-focus X-ray diagnostic neonatal complex".

References

[1] Potrakhov N N and Gryaznov A Yu 2012 *Microfocus X-ray radiography in medical diagnostics* (Publishing house SPbETU "LETI")
[2] Potrakhov E N 2012 *Medical technics* **5** 44–7
[3] Mazurov A I and Potrakhov N N 2014 *Biotekhnosphera* **3** 26–8
[4] Potrakhov E N 2012 *Medical technics* **5** 37–40

Carbiding of the electrodes of electrovacuum devices

E D Pryalukhin[1], A A Rikov[2], D K Kostrin[2] and A A Lisenkov[2]

[1] JSC "SED-SPb" (Svetlana Electron Devices – Saint-Petersburg), Engels pr. 27, Saint-Petersburg, Russia, 194156
[2] Saint-Petersburg State Electrotechnical University, Prof. Popova str. 5, Saint-Petersburg, Russia, 197376

E-mail: lisran@yandex.ru

Abstract. Modification of the surface properties of the materials of electrodes or deposition of protective coatings on them are the most effective ways to increase the service life and reliability of electrovacuum devices. Very often for these tasks operations of the carbiding of details are used. In this work results of direct carbiding of W and Mo, carried out in a mixture of the hydrogen stream and the carbon-containing gas on a technological installation of a flow type, are discussed.

Operational properties of parts and components used in the manufacture of electrovacuum devices are determined by the condition of their surface layer. This is due to the fact that the surface layer of the component during its operation is subjected to the external active mechanical, thermal, chemical or other types of impact. Therefore, modification of the surface properties of the material or deposition of a protective coating are the most effective ways to increase the service life and reliability of parts and devices [1–4].

The most widely as wear resistant coatings the refractory compounds of *d*-transition metals of IV–VI groups of the Periodic system of elements with O_2, C, N_2 are applied, due to the peculiarities of their crystallochemical structure [5].

1. Metals, having a lack of electrons in the inner *s*-, *p*- and *d*-orbitals, quite easily gain electrons from any source, as which can serve internode atoms of carbon, nitrogen and oxygen.

2. Most transition metals have sufficiently large atomic radii and during the formation of compounds with atoms O, C and N_2 can obey the Hagg rule, according to which the ratio of the radius of the nonmetal atom to the radius of the metal atom is less than the critical value of 0.59. For the compounds of metals of IV group (Ti, Zr, Hf) Hagg rule is accurately obeyed, that leads to the formation of simple structures in which prevails the connection metal-to-metal, and atoms C and O can be regarded as inserted into the lattice of metal atoms.

3. Most transition metals have a wide region of homogeneity that allows depending on the content of oxygen, nitrogen and carbon to change the physical and mechanical properties of their carbides, nitrides and oxides quite noticeably.

4. Transition metals and their compounds, primarily compounds with a simple cubic structure of NaCl type (ZrC, ZrN, TiN, VC, TaC), have extremely high melting points.

Very often in the manufacture of electrovacuum devices operations of the carbiding of details are used. So the source of the flow of electrons in vacuum devices is a thermionic cathode, which should ensure high durability and stable emission at the smallest possible consumption of energy for its heating [6]. In addition, the cathode should be resistant to repeated cyclic stress and the working surface of the cathode must not break down when exposed to ions rapidly moving towards the cathode. At the same time, the higher the anode voltage is, the more energy have the ions when they reach the cathode.

During the manufacture of (W–ThO_2–WC) cathodes technological operation of the direct carbiding of the tungsten wire is used: tungsten doped with 1.5...2.0 % ThO_2 or tungsten with additives of 10 % Re and 1.5...2.0 % ThO_2 [7].

The shielding grid is used to improve current takeoff, it is electrically connected to the positive pole of the anode power source and decreases the likelihood of parasitic oscillations at high frequencies by reducing the capacitance between the control grid of the lamp and its anode (figure 1). To increase the service life of devices antiemission coatings are applied [8–10], to improve the adhesion of which, operations of carbiding of the surface of the mesh fabric are also used. To obtain a layer of the substrate material carbide, was used as a technological devices, providing a direct carbiding in an atmosphere of carbonaceous gas, and also vacuum arc plasma sources with integrally cold cathode [11].

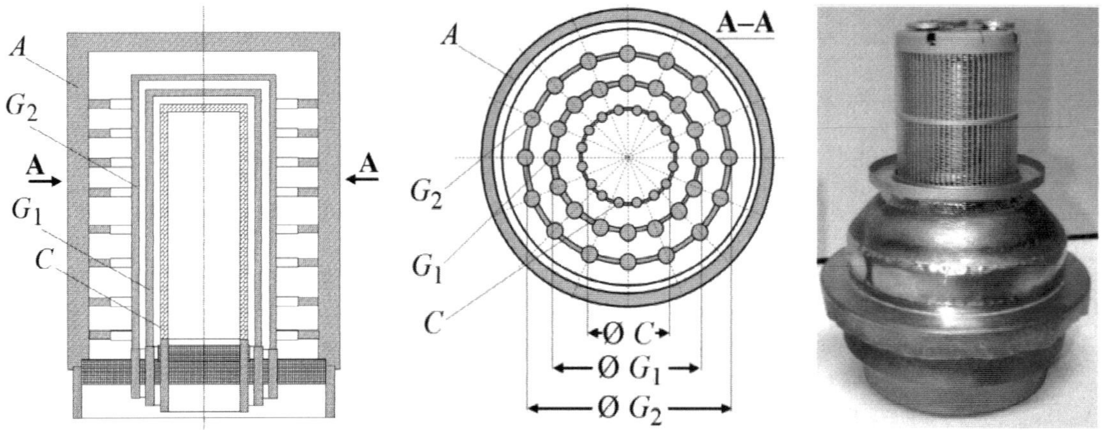

Figure 1. Generator lamp: **(a)** – design (*C* – cathode; G_1 – control grid; G_2 – screen (shielding) grid; *A* – anode); **(b)** – assembly of a cathode-grid node of a GU-92B lamp

The phase composition of the formed coatings was studied by X-ray diffraction analysis. The elemental composition of the coatings was determined by electron probe microanalysis, based on comparing the characteristic X-ray spectra of the analyzed sample with standards of known composition. Direct carbiding of W and Mo (heavy transition elements of the sixth group with very similar mechanical, physical and chemical properties) [12] was carried out in a mixture of the hydrogen stream and the carbon-containing gas on a technological installation of a flow type.

To exclude the possibility of oxidation of the substrates and their decarbiding in the process of heat treatment the hydrogen was carefully purified from oxygen and water vapors. During annealing of the substrate to a temperature of 1500 K and above in the region adjacent to its surface, the thermal decomposition of hydrocarbons with the formation of the atomic carbon occurs, forming carbide compounds [13], while gaseous components are removed from the chamber and neutralized. The optimum condition for the formation of carbide phases and absence on the surface and in the bulk of the liquations of free carbon, is the achievement of the balance between the speed of the process of formation of atomic carbon and its rate of diffusion into the material [14, 15].

When determining the thickness of the carbide layer the fact that one of the characteristics of the resulting carbides is emerging fragility of the material, increasing with the growth of the carbide layer, was taken into account. The thickness of the carbide layer was evaluated by measuring the electrical resistance before and after the process of carbiding and by studying the metallographic sections.

As the result of carbiding the resistance of a material changes, that allows to assess the degree of carbiding γ using a method of measuring the electrical resistance of the metal before carbiding R_{Me}

and after the process of carbiding R_{MeC}: $\gamma = [\rho_{MeC} / (\rho_{MeC} - \rho_{Me})][\gamma' / (1+\gamma')]$, where ρ_{MeC} and ρ_{Me} – resistivity of carbide and pure metal, respectively; $\gamma' = (R_{MeC} - R_{Me}) / R_{Me}$ [16].

The advantage of using vacuum arc plasma sources is an opportunity to consistently obtain in a single technological cycle of the substrate material carbide (MoC, WC) and deposit coating of zirconium carbide. In this case, the formation of the coating is realized by deposition of ions with high kinetic energy. Changing the value of the accelerating potential U allows to adjust the energy of the deposited ions, and hence to control the flow of the process of the coating formation. Preliminary condensation of the sprayed cathode material (Mo, W) to a face of an akin metal contributes to achievement of a structural-based coating, which in a subsequent diffusion reaction reduces the interfacial tension related to the difference of the crystal lattices, and improves the strength of the adhesive bond. For the implementation of plasma-chemical synthesis of carbide compounds into the flow of metal plasma of a vacuum arc discharge reactive gas benzene (C_6H_6) is introduced.

Mo has the atomic radius 1.386 Å and W – 1.394 Å, but their atomic masses (m_W = 183.8, m_{Mo} = 95.94) differ in m_W/m_{Mo} = 1.916 times, which leads to significantly different processes of interaction with plasma flow. In addition, because of the peculiarities of the electronic structure and complex covalent-ionic-metallic nature of the interatomic interactions the phase diagrams of the carbide systems of Mo and W have several structural modifications, stable in different temperature and concentration intervals. For the selected conditions of carbiding were obtained layers consisting, respectively, of the phases of molybdenum carbide – MoC and tungsten carbide, formed from the inner layer of halfcarbide W_2C and surface monocarbide layer of WC with a hexagonal structure. With the increase in time of carbiding the thickness of the W_2C layer increases, while the thickness of the surface WC layer has changed very little. Growth of the diffuse layers obeys parabolic temporary law [17], typical for the processes of reaction diffusion, which indicates the preferential diffusion of atoms of the nonmetal – carbon, through the lattice of the formed phases into the metal.

For the formed coatings were observed all the lines of maximum intensity characteristic to these carbides. So for lines of W_2C should be noted: (321, 302) – 1.3471 (*ASTM* – 1.348) the intensity of the line 40; (200) – 2.3584 (*ASTM* – 2.363) the intensity of the line 35; (221, 202) – 1.7454 (*ASTM* – 1.749) the intensity of the line 35; (040, 023) – 1.4955 (*ASTM* – 1.502) the intensity of the line 30. In the coating there is WC compound with line: (302, 116) – 1.2836 (*ASTM* – 1.281) the intensity of the line 80. The presence in the coating of the various phases of tungsten carbide is due to the complexity of its formation at these temperatures.

The presence of the sublayer of substrate material carbide $Me_{subst}C$ with thickness up to 1 μm provides during the subsequent formation from the flow of the metal plasma of vacuum arc discharge in the environment of carbon-containing gas achievement of a coating of zirconium carbide ZrC with stable composition. After the deposition in the electrolytic solution of platinum layer during heat treatment in vacuum, in the formed system $Me_{subst}(Mo,W)–Me_{subst}C–ZrC–Pt$, occurs the formation of an intermetallic compound Pt_3Zr.

References

[1] Kostrin D K and Lisenkov A A 2016 *Materials Science Forum* **843** 278–83
[2] Vinogradov M L, Kostrin D K, Smelova V V, Trifonov S A and Lisenkov A A 2016 *Proceedings of the 2016 IEEE North West Russia Section Young Researchers in Electrical and Electronic Engineering Conference* 729–30
[3] Kostrin D K and Lisenkov A A 2017 *Vakuum in Forschung und Praxis* **3** 35–9
[4] Kostrin D K, Lisenkov A A and Potrakhov N N 2017 *Biomedical Engineering* **4** 262–6
[5] Kostrin D K and Lisenkov A A 2016 *Materials Science Forum* **870** 371–6
[6] Beck A H W 1953 *Thermionic Valves: Their Theory and Design* (London: Cambridge University Press)
[7] Serova N N and Lesish Yu K 1967 *To the question of durability of tungsten thoriated and carbonized cathode, Electronic instrumentation* (Moscow–Leningrad: Energy) 124–7

[8] Bystrov Yu A, et al. 2009 *Modern technological processes in the production of powerful generator lamps* (Saint Petersburg: Publishing House of Saint Petersburg Electrotechnical University "LETI")

[9] Vetrov N Z, Kostrin D K, Lisenkov A A and Popova M S 2015 *Journal of Physics: Conference Series* **652** 012032

[10] Lisenkov A A, Vetrov N Z and Kostrin D K 2017 *Technical Physics Letters* **4** 390–2

[11] Bystrov Yu A, Vetrov N Z, Lisenkov A A and Kostrin D K 2014 *Vakuum in Forschung und Praxis* **5** 19–23

[12] Kubo R 1970 *Thermodynamics* (Moscow: Mir)

[13] Kostrin D K, Pikus M I, Trifonov S A and Lisenkov A A 2017 *Journal of Physics: Conference Series* **789** 012029

[14] Lisenkov A A, Kostrin D K and Pikus M I 2017 *Solid State Phenomena* **265** 750–4

[15] Pikus M I, Lisenkov A A, Kostrin D K and Trifonov S A 2017 *Journal of Physics: Conference Series* **929** 012093

[16] Prilutskiy V S 2001 *Tungsten thoriated and carbonized cathode* (Moscow: Ore and metals)

[17] Gurov K P, Kartashkin B A and Ugaste Yu E 1981 *Mutual diffusion in multiphase metallic systems* (Moscow: Nauka)

IOP Conf. Series: Journal of Physics: Conf. Series **967** (2018) 012014 doi:10.1088/1742-6596/967/1/012014

Quantitative data standardization of X-ray based densitometry methods

K A Sergunova[1]*, A V Petraikin[1], F A Petrjajkin[1], K S Akhmad[1], D S Semenov[1] and N N Potrakhov[2]

[1] Research and Practical Center of Medical Radiology, Department of Health Care of Moscow, 109029, Moscow, Russia
[2] Saint Petersburg Electrotechnical University "LETI", 197376, St. Petersburg, Russia

E-mail: ska@rpcmr.org.ru*

Abstract. In the present work is proposed the design of special liquid phantom for assessing the accuracy of quantitative densitometric data. Also are represented the dependencies between the measured bone mineral density values and the given values for different X-ray based densitometry techniques. Shown linear graphs make it possible to introduce correction factors to increase the accuracy of BMD measurement by QCT, DXA and DECT methods, and to use them for standardization and comparison of measurements.

Today the diagnosis of osteoporosis (OP) is a relevant issue in Russia and all over the world. Population studies indicate that every third woman and one of four men over 50 years old suffer from OP. More than 40 % of people of both sexes have signs of osteopenia.

Bone mineral density (BMD) is a quantitative parameter used to estimate bone strength and mechanical properties and, therefore, to determine the risk of fracture. BMD measurement is performed using one of the following X-ray methods: dual-energy X-ray absorptiometry (DXA), quantitative computed tomography (QCT), and a new type of computed tomography (CT) – dual-energy computed tomography (DECT).

The wide dissemination of X-ray based densitometry methods both for the diagnosis of OP and for the control purpose during treatment requires the accuracy and reproducibility of BMD measurements. This leads to the need for standardization of BMD determining methods, developing control methods for cross-calibration, which allows comparing the results of the determination of BMD on different devices and different methods [1].

There are cases of inaccuracies in determination of BMD for DXA associated with the type of used equipment, calibration procedure, as well as anatomical features of the patient [1, 2].

The European Spine phantom (ESP) had been developed in 90s to conduct daily quality control and cross-calibration between different devices [1]. It consists of three sections, each of which has the form of a vertebra with different content of bone minerals. This design solution allowed the use of a standard patient scan protocol for both methods: DXA and QCT.

The ESP phantom was used to compare DXA scanners before and after standardization of measurements [1, 4]. The significant differences in the readings of the scanners and the given BMD, as well as the results of comparison of apparatuses with each other, confirmed the need for standardization of measurements during densitometry. The comparable BMD measurement results were obtained only after the corresponding adjustments. The use of ESP phantom not only increased the quality of routine diagnostic procedures, but also conducted multicenter clinical studies using cross-calibration procedure.

However, CT image errors from beam hardening and X-ray scatter causing inaccuracy of CT numbers, contrast reduction, and cupping artifacts should be taken into account [3] The ESP phantom does not make it possible to move objects freely, simulating different vertebrae position in the plane of the phantom section, since it is filled with epoxy rubber. Thus, it could not be used for modeling of different X-ray density along the cross section of a cylindrical test object. In addition, composites used in the manufacture as an equivalent of water [1] can cause inaccuracies in the decomposition of BMD at DECT.

In order to overcome these, as well as to perform a comparison of various X-ray methods for determining BMD, a special liquid phantom (SL-phantom) was developed. The SL-phantom has 4 cylindrical plastic "vertebras" bodies (diameter and height 3.5 cm, wall thickness 2 mm), manufactured by the method of three-dimensional printing (figure 1(a)). Cylinders were filled with solutions of dipotassium phosphate of various concentrations. This substance is highly soluble and allows to simulate the BMD in a wide range. A bracket of radiopaque material was designed to provide the "vertebrae" positioning. The phantom was immersed in a tank with a physiological solution simulating a human body with a diameter of 32 cm. The vertebrae could move in the bracket, occupying positions in the center of the cylinder with liquid and around the periphery (12 cm from the center). Dipotassium phosphate concentrations were: 29.4 and 60.3 mg/cm^3, simulating OP; 92.6 mg/cm^3 for simulating osteopenia; 161.4 mg/cm^3 for simulating normal BMD values. Threshold values were chosen based on the recommendation of the WHO and ACR.

The CT images of the phantom at the center (figure 1(b)) and at the periphery of the tank with saline were processed using the QCT-program to obtain the mineral density value. CT study was performed on a 64-row CT-scanner (120 kV, 100 mA, FOV 40 cm, FC08 core), software for determining the BMD (QCT with an asynchronous phantom). DXA was performed on X-ray densitometer. Digital radiography was performed on a routing X-ray DR-scanner. DECT with densitometry was performed on a multirow CT-scanner with a tube voltage modulation (80–140 kV) (figures 2(a)–2(d)).

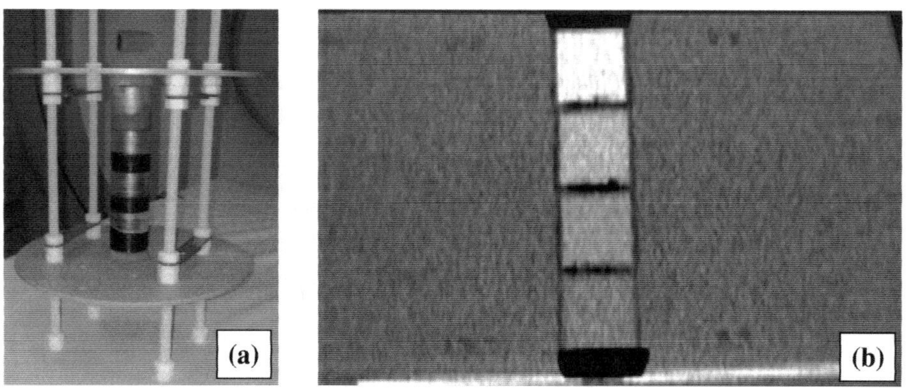

Figure 1. SL-phantom: **(a)** – appearance; **(b)** – CT-image.

In assessing the accuracy of quantitative data, the QCT method was used to determine the linear dependence between BMD and the X-ray density values in the investigated range (see figure 2(a)). However, when the phantom is located in the center of the tank simulating the human body, a systematic overestimation in the density by approximately 10 mg/cm^3 was observed, which corresponds to an error of up to 34 % for the "vertebra" of the lowest density. The error in measuring the phantom at the periphery of the container did not exceed 6.6 %. Taking into account the strictly linear dependence of the obtained curves (the determination coefficient was $R^2 = 1.0$), these errors can be eliminated by introducing correction coefficients for the corresponding phantom arrangements.

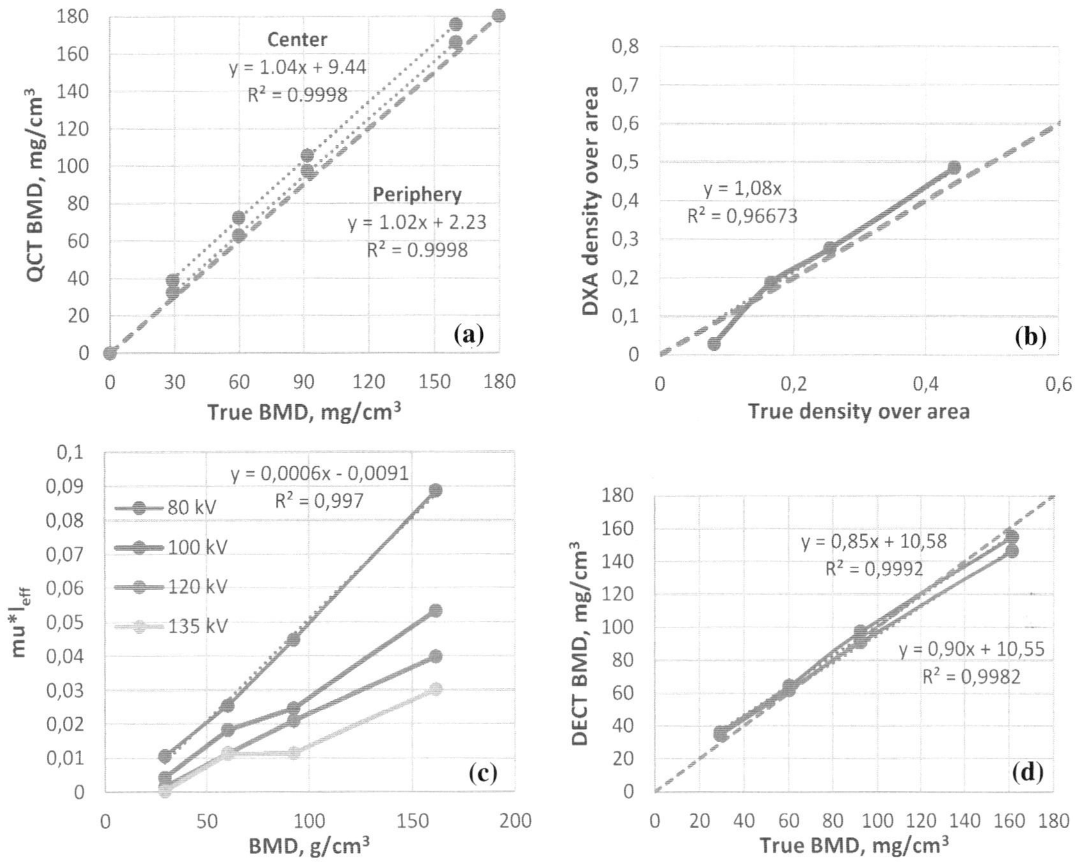

Figure 2. Dependencies between the measured bone mineral density values and the given values for different X-ray based densitometry techniques: **(a)** – QCT; **(b)** – DXA; **(c)** – digital radiography; **(d)** – DECT (in (a), (d) – blue in center, orange – in periphery).

The results of the study of the phantom by X-ray radiography are shown in figure 2(c). Linear dependence was established at a voltage of 80 kV. Distortions in the shape of the curve and the angle of inclination were observed when choosing other values of the anode voltage. Thus, it is not possible to select a single calibration curve, which makes it impossible to consider this method as promising taking into account the automatic shooting mode.

When scanning the SL-phantom by the DXA method, a linear dependence of the measured BMD values on the set values was established (see figure 2(b)), $R^2 = 0.98$. The correct results were obtained with a decrease in the liquid layer to 15 cm. The error of the DXA method in absolute value is insignificant; however, in the region of low BMD values (cortical mineral density was not taken into account) it can be up to 65.4 %. Thus, the parameters of the patient's body significantly affect the results of DXA, and in the low-density region the error of this method may increase.

In contrast to the QCT method, the dependency graphs of the DECT BMD have different slope coefficients when positioned at the periphery and in the center (see figure 2(d)). DECT study demonstrated a linear dependence ($R^2 = 0.99$) between determined potassium concentration and its true both in the center of the container and on the periphery without using a special calibration phantom. It was established that the DECT method is the least sensitive to the shape of the human body, and the error of this method does not have a clear dependence on the true BMD. However, in the range of values corresponding to osteoporosis, it is quite significant (up to 21.9 %).

It is worth noting that the shown linear dependencies make it possible to introduce correction factors to increase the accuracy of BMD measurement by QCT, DXA and DECT methods, and to use them for standardization and comparison of measurements. Thus, an increase in the number of patients who have a high risk of osteoporotic fractures leads to the need for timely diagnosis using modern X-ray based densitometry methods: QCT, which allows to measure BMD not only from the healthcare organization's CT data, but also by opportunistic screening, DECT and DXA.

To ensure the accuracy and reproducibility of BMD measurements, it is necessary to develop control techniques that are carried out with the help of appropriate phantoms. The creation and implementation of the standardization of densitometric studies will allow to assess the reliability of the research conducted on various scanners, to determine the factors that affect the accuracy of measurements, and to make appropriate corrections. In this paper, a phantom prototype is presented. It was used for BMD measurement errors estimation for various X-ray based densitometry techniques.

References

[1] Kalender W A, Felsenberg D, Genant H, Fischer M, Dequeker J and Reeve J 1995 *European J Radiology* **20** 83–92

[2] Novikov V E, Skripnikova I A, Murashko L M and Abirova E C 2014 *Osteoporosis and osteopathy* **1** 39–42

[3] Engelke K, Lang T, Khosla S, Qin L, Zysset P, Leslie W D, Shepherd J A and Shousboe J T 2015 *J Clin Densitom* **18(3)** 393–407

[4] Park A J, Choi J H, Kang H, Park K J, Kim H Y, Kim S H, Kim D Y, Park S H and Ha Y C 2015 *J Bone Metab* **22(2)** 45–9

Application of the gas-discharge surge arresters in X-ray devices and low voltage instrumentation

V A Simon, V A Gerasimov, D K Kostrin, A A Lisenkov, L M Selivanov and A A Uhov

Department of electronic instruments and devices, Saint Petersburg Electrotechnical University "LETI", 197376, Saint Petersburg, Russia

E-mail: vsev.simon@gmail.com

Abstract. Usage of the gas discharge in science and engineering is discussed. Application examples of the compact gas-discharge tubes in the X-ray devices and low voltage instrumentation appliances for the surge protection are presented.

1. Introduction

The electrical discharge in gases is widely used in science and in leading branches of industry. The partial list of the application fields where gas discharge brings benefits includes optics, metrology, materials science, electronics, radio engineering, mechanical engineering, healthcare and illumination.

For instance, fluorescent lamps utilize ultraviolet light of the discharge in mercury vapors [1] which excites the phosphor of the lamp and thus engenders the white light. Mercury-vapor lamps with fused quartz casing don't have any phosphor and used for ultraviolet disinfection of the medical chambers, kindergarten and school rooms, also for bactericidal cleaning in tap water. High-pressure sodium lamps [2] are implemented worldwide as a reliable and energy-efficient light source for road lighting and as plant grow light for the agricultural and decorative plants. The gas discharge is also necessary for the operation of the lasers with gas working medium [3]: carbon dioxide [4], helium-neon, helium-cadmium and other suchlike gas mixtures.

The magnetron discharge systems with high pulse current emitted from the cold cathode are capable to create intensive electron beams [5]. Magnetron sputtering of the metal targets [6] is widely applied for thin film growing. The cathode sputtering in vacuum chamber by means of the arc discharge allows formation of the coatings with various chemical compositions [7–16] and cleaning of the dielectric surfaces with the flow of neutral particles [17]. The spectra of electric discharge with the liquid nonmetallic cathode [18, 19] can give useful information related to the dangerous impurities (e.g. heavy metals) existing in the liquid.

2. Surge protectors

It is usually required in electrical and electronic systems to avoid the arising of the high-voltage pulses in the electronic components designed for the low-voltage signals. This requirement can be satisfied using the surge protectors (also referred as surge arresters): metal oxide varistors, transient voltage suppressors, thyristor surge protection devices, air spark gaps and gas-discharge tubes. Each type of the listed surge protectors has its own properties fit for the certain exploitation conditions.

IOP Conf. Series: Journal of Physics: Conf. Series **967** (2018) 012015 doi:10.1088/1742-6596/967/1/012015

The surge arresters in the form of the gas-discharge tube have negligible small capacitance (often less than 1 pF) and can handle relative high currents. In case of the high-voltage pulse the surge arrester creates short circuit between its pins thus dissipating the excessive energy. The gas discharge tubes are manufactured in the cylindrical enclosure for through-hole mounting with large metal leads as well as in the compact surface mount (SMD) casing. The last is important for use in portable electronics such as radio receivers, walkie-talkies, MP3 audio players, mobile phones, smart watches, small wearable Holter monitors, electroencephalographs or pulse oximeters.

3. Gas-discharge surge protectors in instrumentation

The low-power glow discharge thyratrone MTH-90 (figure 1) can be used as a gas plasma arrester in instrumentation. The discharge ignition voltage between the cathode and grid of the MTH-90 is over the range from 65 to 90 V. This feature is applicable for the X-ray tube current I_{XRT} measurement by means of the resistive shunt R_{shunt} (10 Ω) to protect the instrumentation electronics from failure due to high-voltage pulses (figure 1(c)). In addition to the thyratrone the resistive current limiters R_{lim} (100 Ω) and the capacitive voltage limiter C_{lim} (470 nF) are used.

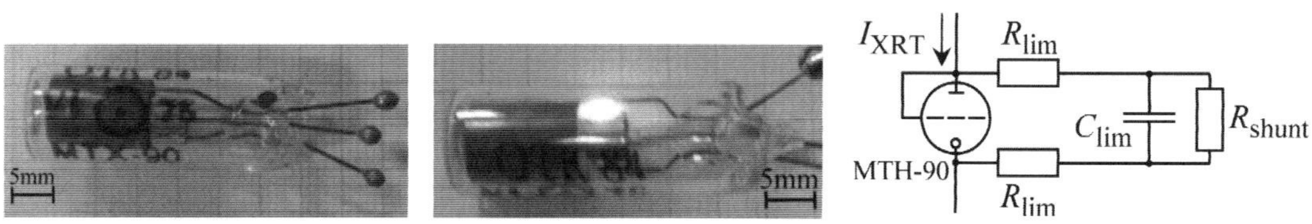

(a) **(b)** **(c)**

Figure 1. Appearance of the MTH-90 thyratrone **(a)**, light of the glow discharge **(b)**, circuit for the surge protection of the shunt resistor and additional instrumentation electronics **(c)**.

The modern gas-discharge surge arresters (figure 2) are nontransparent gas-filled tubes with metal or ceramic shell with two or three leads. One of the widespread examples of the general-purpose surge protectors is the EC90X unit manufactured by Epcos (part of the TDK Group). This device has two long axial leads and industry-standard tubular shape. The EC90X arrester has following electrical specifications: the direct current spark-over voltage is 90 V, the peak pulse current is 10 kA, and the insulation resistance is at least 1 GΩ [20]. Another widely distributed arrester is the Epcos S30-A230X unit in space-saving SMD casing; this device also features high insulation resistance (>1 GΩ), spark-over at 230 V (direct current) and peak pulse current of 1 kA [21].

The S30-A230X surge arresters were used for the electrostatic protection of the INA333 instrumentation amplifier (IA) [22]. Two arresters were connected between inputs of the IA and the ground circuit. Two 100 MΩ pull-down resistors R_{bypass} were plugged in parallel with the arresters for maintenance of sufficient value of the input bias current for the IA (figure 2(b)). In order to test the functioning of the protection the 6 kV pulse generator was applied. The IA has successfully continued the functioning after the surge test which indicates the high effectiveness of the protection.

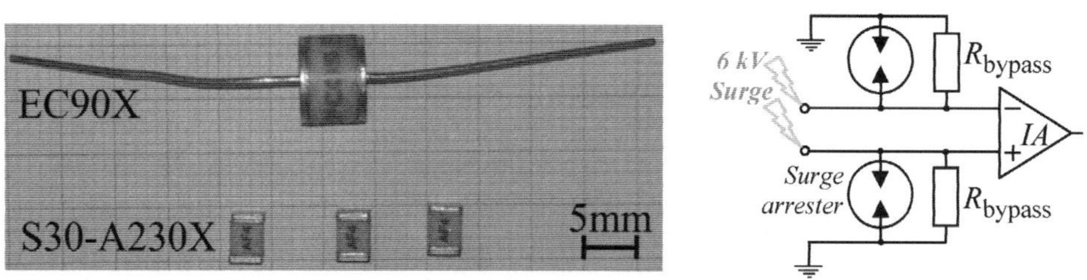

(a) **(b)**

Figure 2. Appearance of the EC90X and S30-A230X arresters **(a)**, connection of the S30-A230X surge arresters for the IA protection **(b)**.

4. Conclusion

Described technique of the surge protectors application is suitable for low voltage instrumentation such as precision current sensing, bioelectric activity registration and Wheatstone bridge based measurements due to high insulation resistance and very low capacitance of the gas-discharge arresters.

Acknowledgement

Described results were obtained during the research and development efforts in accordance with the 218th Ordinance of the Russian Federation, treaty № 02.G25.31.0192.

References

[1] Perkin F M 1911 *Trans. Faraday Soc.* **6** 199–204
[2] de Groot J J and van Vliet J A J M 1986 *The High-Pressure Sodium Lamp* (Deventer: Kluwer Technische Boeken BV)
[3] Kiselev A S, Kostrin D K, Lisenkov A A and Smirnov E A 2017 *J. Phys.: Conf. Ser.* **789** 012027
[4] Kostrin D K, Chernigovskiy V V, Martsinukov S A and Lisenkov A A 2016 *Vakuum in Forschung und Praxis* **6** 34–7
[5] Udovichenko S Yu, Kostrin D K and Lisenkov A A 2017 *J. Phys.: Conf. Ser.* **872** 012013
[6] Gerasimov V A, Kostrin D K, Selivanov L M, Uhov A A, Komlev A E and Shutova E S 2016 *J. Phys.: Conf. Ser.* **729** 012019
[7] Bystrov Yu A, Kostrin D K, Vetrov N Z and Lisenkov A A 2014 *Vakuum in Forschung und Praxis* **5** 19–23
[8] Bystrov Y A, Kostrin D K, Lisenkov A A and Vetrov N Z 2015 *Vakuum in Forschung und Praxis* **2** 22–5
[9] Kostrin D K and Lisenkov A A 2016 *Materials Science Forum* **843** 278–83
[10] Kostrin D K, Uhov A A and Lisenkov A A 2016 *Vakuum in Forschung und Praxis* **3** 34–7
[11] Kostrin D K, Uhov A A and Lisenkov A A 2016 *J. Phys.: Conf. Ser.* **735** 012055
[12] Kostrin D K and Lisenkov A A 2016 *Materials Science Forum* **870** 371–6
[13] Lisenkov A A, Vetrov N Z and Kostrin D K 2017 *Technical Physics Letters* **4** 390–2
[14] Kostrin D K and Lisenkov A A 2017 *Vakuum in Forschung und Praxis* **3** 35–9
[15] Kostrin D K, Lisenkov A A and Potrakhov N N 2017 *Biomedical Engineering* **4** 262–6
[16] Lisenkov A A, Kostrin D K and Pikus M I 2017 *Solid State Phenomena* **265** 750–4
[17] Fadeev A S, Talanov A S, Kostrin D K, Trifonov S A, Lisenkov A A and Goncharov V D 2017 *J. Phys.: Conf. Ser.* **872** 012021
[18] Ramazanov A N, Kostrin D K, Simon V A, Uhov A A and Lisenkov A A 2017 *Proc. of the 2017 IEEE Conf. of Russian Young Researchers in Electrical and Electronic Engineering (EIConRus)* (Saint Petersburg) pp 58–62
[19] Ramazanov A N, Kostrin D K and Lisenkov A A 2017 *J. Phys.: Conf. Ser.* **929** 012062
[20] *B88069X0720S102 Surge Arrester (EC90X 2-electrode Arrester)* URL: https://en.tdk.eu/inf/100/ds/ec90x_x0720.pdf
[21] *B88069X5941T203 Surge Arrester (S30-A230X 2-electrode Arrester)* URL: https://en.tdk.eu/inf/100/ds/s30a230x_x5941.pdf
[22] *INA333 Micro-Power (50μA), Zero-Drift, Rail-to-Rail Out Instrumentation Amplifier.* URL: http://www.ti.com/lit/ds/symlink/ina333.pdf

Microelectronic devices digital X-ray image processing method development

N E Staroverov, A Yu Gryaznov, E D Kholopova and K K Guk

Department of electronic instruments and devices, Saint Petersburg Electrotechnical University "LETI", 197376, Saint Petersburg, Russia

E-mail: nik0205st@mail.ru

Abstract. In this paper microelectronic devices digital X-ray image processing method development is described. The main steps of the algorithm work are presented. The results of using the algorithm for improving the printed circuit board image are shown

Nowadays the X-ray digital image processing in nondestructive testing and medicine diagnostic are widely used. The methods of X-ray digital image processing allow cleaning images from noise, changing the brightness and contrast of images, improving the discernibility of small details on the image. Thus digital image processing can increase the amount of information in the X-ray image. In some cases, digital image processing can compensate for the unevenness of the X-ray intensity along the image field or incorrectly selected modes of operation of the X-ray apparatus. In this paper, the method for improving the quality of X-ray images of printed circuit board (PCB) is proposed. The algorithm for increasing the information content of PCB images consists of the following stages:

– image acquisition;
– image background correction;
– contrasting the image;
– calculation of the global threshold, threshold transformation of the image by the Otsu method;
– application of morphological filtration for the selection of interesting areas of the image;
– two-dimensional fast Fourier transform of image, suppression of low frequencies of image spectrum, inverse fast Fourier transform;
– combining images obtained by automatic contrast and high-frequency filtering, obtaining the final result.

To allocate borders on the image, an image segmentation method based on the global threshold was applied. The global threshold was determined by the Otsu method. The Otsu method only uses an image histogram, so it takes much less time to compute in comparison with methods based on image analysis. The main idea of Otsu method is the fact that if the pixel classes on the image are well separated in brightness, that they are as well separated on the threshold. The threshold will be optimal if the variance between the two classes of pixels to which the image has divided the threshold, will be maximum. The threshold is to be found by searching all possible integer values for the brightness of the image (in the case of a jpg or bmp image, 256 values need to be sorted).

Figure 1 shows the original image and the result of the threshold transformation described by the expression (1), where k is the threshold value

$$f(x,y)=\begin{cases}1, & g(x,y)>k \\ 0, & g(x,y)<k\end{cases}. \tag{1}$$

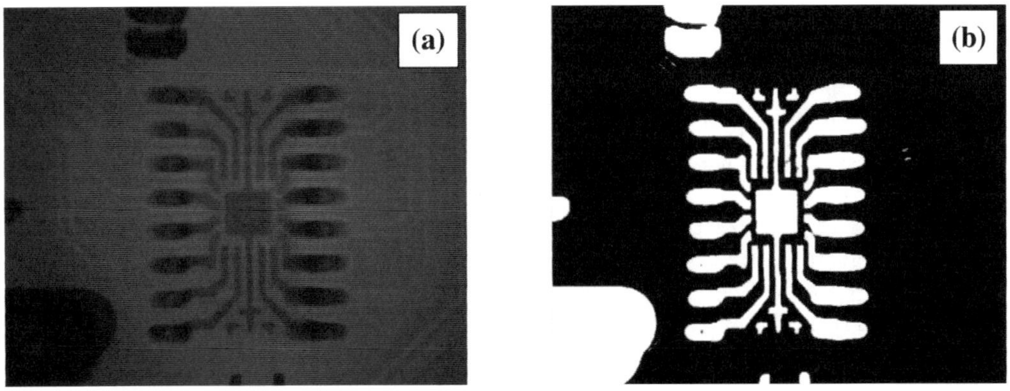

Figure 1. Original image **(a)** and segmentation of the image based on the global threshold **(b)**.

As can be seen in figure 1, the threshold transformation quite well selects the object (in this case, the microcircuit), but several sections of the object still remain unselected (black areas in the image of the chip). To eliminate the considered disadvantages of threshold processing, morphological processing of the image was applied. To isolate the remaining areas, morphological expansion and morphological erosion operations were consistently applied. The image after applying the morphological processing is shown in figure 2. As can be seen from the figure, the object in the image is completely white, there are no isolated groups of black pixels. Isolation of the chip in the image was necessary to change the contrast and increase the sharpness of only the image of the chip with the same background of the entire image.

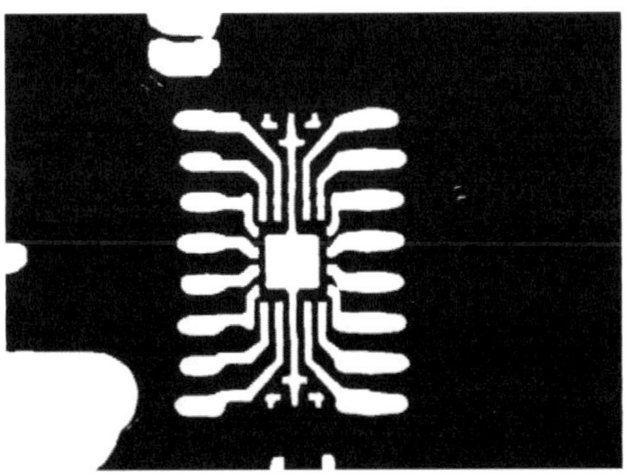

Figure 2. Image after morphological processing.

To enhance the contours in the image and also to increase the contrast, filtering in the frequency region was applied. The two-dimensional Fourier transformation and the inverse transformation were carried out according to formulas (2) and (3)

$$F(u,v)=\frac{1}{MN}\sum_{x=0}^{M-1}\sum_{y=0}^{N-1} f(x,y)e^{-i2\pi(ux/M+vy/N)}; \tag{2}$$

$$f(x,y)=\frac{1}{MN}\sum_{u=0}^{M-1}\sum_{v=0}^{N-1} F(u,v)e^{i2\pi(ux/M+vy/N)}. \tag{3}$$

The filtering was carried out using a high-pass Gaussian filter, which attenuates low frequencies while the rest remains unchanged. Then the inverse Fourier transformation was carried out. The result of applying high-frequency filtering to the test image is shown in figure 3.

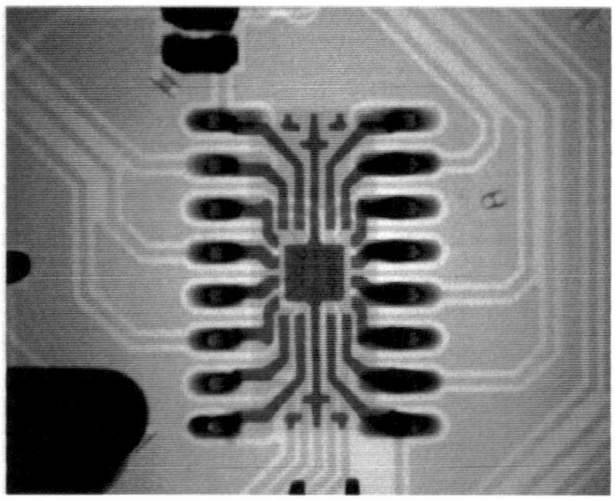

Figure 3. The result of filtering in the frequency region.

The last step of the algorithm was the combination of images obtained by high-frequency filtration and obtained by contrasting. The image of the chip, obtained by the frequency filtering method, was superimposed on the background image obtained by auto-contrast of the original image, the superposition was carried out using an image obtained by the threshold transformation. If a pixel had a value of 1 on the binary image the corresponding pixel on the image was replaced by the pixel from the image obtained by high-frequency filtration.

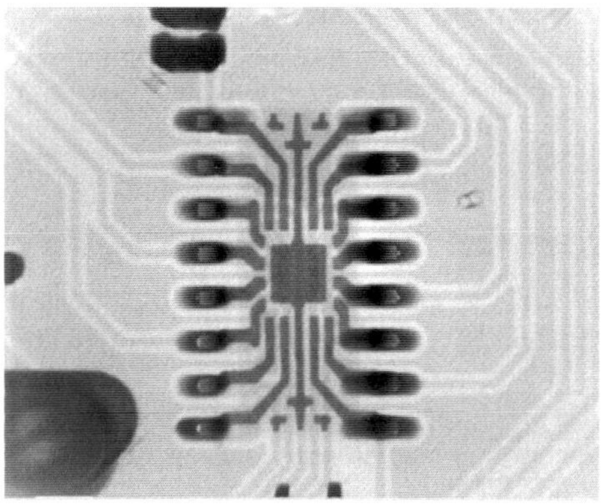

Figure 4. The image obtained as a result of application of the developed algorithm

The result of algorithm work is shown in figure 4. In comparison with the original image, the contrast of the defects increased and the sharpness of the contours in the image increased significantly.

References

[1] Potrakhov N N 2007 *Journal of new medical technologies* **3** 167–9
[2] Potrakhov N N, Gryaznov A Yu, Zhamova K K, Bessonov V B, Obodovski A V, Staroverov N E and Kholopova E D 2015 *Biotechnosphere* **5** 46–54
[3] Gryaznov A Yu 2010 *Methods and systems microfocus phasecontrast x-ray medical. Doct. Diss.* (Saint Petersurg)
[4] Gonzalez R C and Woods R E 2008 *Digital Image Processing* (Prentice Hall Publ.)

IOP Conf. Series: Journal of Physics: Conf. Series **967** (2018) 012017 doi:10.1088/1742-6596/967/1/012017

Miniature sources of irradiation for intracavitary thermo radiotherapy

M Taubin[1], D Chesnokov[2] and A Simonov[2]

[1] National Research Nuclear University ("MEPHI"), Moscow
[2] Federal State Unitarian Enterprise "Scientific Research Institute Scientific Industrial Association "LUCH", Podolsk, Moscow region, Russian Federation

E-mail: taubin@sialuch.ru

Abstract. This report presents the development of a miniature ionizing and thermal radiation source for oncological diseases treatment namely the inward parts of the body. This source can be placed next to the tumor inside of the body. This report is only about methods and devices for the intracavitary therapy. Irradiation by external sources wasn't considered in our investigation.

Radiologic therapy can be represented as two directions: radiation therapy and hyperthermia (figure 1). In our opinion just these procedures can be the most helpful for oncological diseases treatment. Hyperthermia may be external with body heating or intracavitary with high-frequency sources for tumor overheating (microwave and infrared). Radiation therapy includes brachytherapy and radiotherapy (roentgen therapy). Brachytherapy uses radioactive isotopes either in front of implant or probe with isotopes which placed inside it.

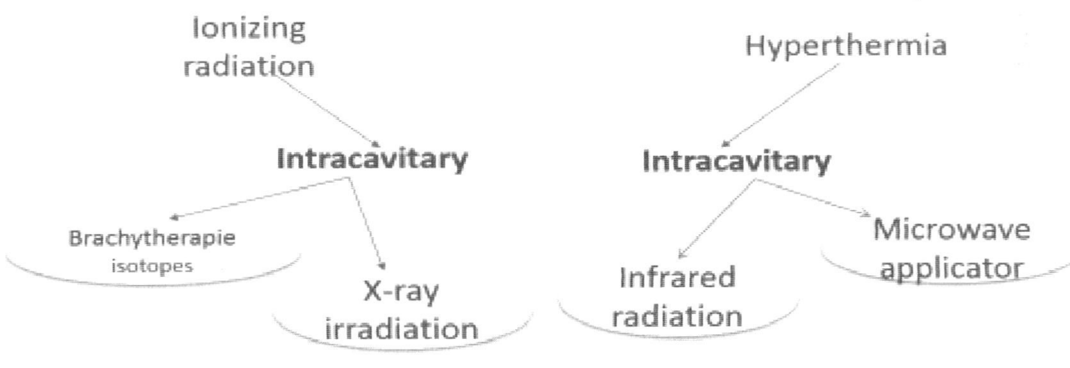

Figure 1. Radiologic therapy.

One direction of Roentgen therapy is X-ray therapy with field emission cathode X-ray tubes. It would like to note that X-ray tubes with field emission cathode with nanotubes is really prospective. In this field our colleagues from the University of North Caroline, Nagoya institute of Technology (Japan), KAIST (Korea) and others have succeeded a lot.

Another direction of Roentgen therapy is use X-ray tubes with thermoemissive cathode. It is known that therapy can be administered with the use of either ionizing radiation or thermal effect (overheating). However, it is mentioned that the treatment is from 1.5 to 2 times more efficient with the use of sequence ionizing and thermal irradiation, but when they are used simultaneously the treatment is from 2 to 4 times more effective, that is the most efficient (figure 2).

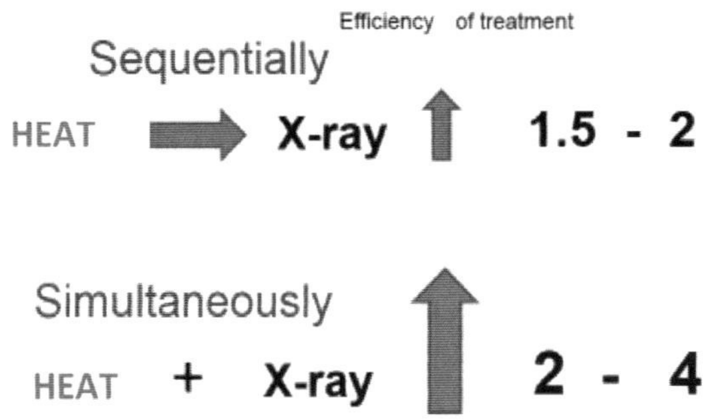

Figure 2. Efficient of treatment by thermoradiation.

Meanwhile the authors of the named studies underline that devices for such treatment are absent. It follows that the problem of making one source allowing irradiating with both X-ray and thermal fluxes is very urgent. It is appropriate to mention one matter of principle here. Size of focal sport required for diagnostics are not needed for therapy in general and for intracavitary therapy in particular. But enhanced power is undoubtedly desirable for reduction of the treatment procedures duration.

Thereby, a source for intracavitary therapy shall fulfil the following requirements: possibility of irradiating with both X-ray and thermal fluxes and enhanced power in spite of the small size on the cathode and anode.

With all the advantages of cold cathodes sources they do not meet these requirements. That means that thermoemissive cathode sources were to be considered. It is known that electron energy is transformed into X-radiation and heat in X-ray tubes with thermoemissive cathodes (figure 3).

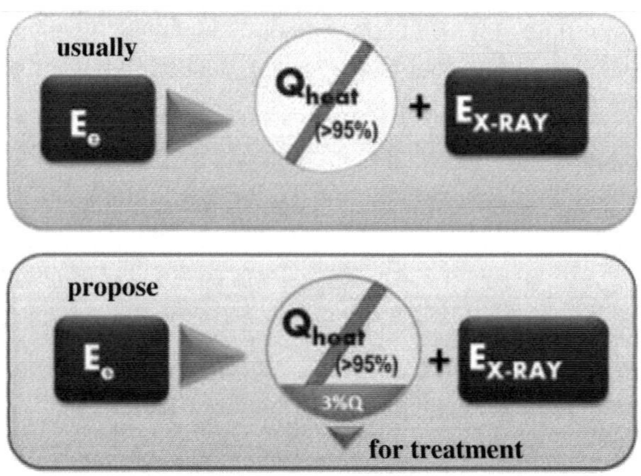

Figure 3. Distribution of electron energy.

At the same time less than 5 % of the energy of the electrons is transformed into X-ray, but more than 95 % is transformed into heat. That is why an X-ray tube should have been taken as the foundation.

But meanwhile it had to solve next questions. How to chill the tube? What materials of a small-sized anode can withstand enhanced specific heat demand? What materials of a small-sized cathode will provide enhanced emission characteristics? How to extract the radiation with the least loss for absorption of gamma-radiation on the way from the anode to the tumor? What material should be used for the envelope? How to provide heat insulation to the lateral surface of the envelope?

Authors of this report have developed such source that meets these requirements. It generates X-radiation and the heat, emitted by electrons slowdown on the X-ray tube anode is used for heating the tumor zone.

A controlled cooling system is developed for that purpose it is based on heat radiation, convection and the use of a heat pipe. There are several models of source design that differ from each other by the cooling systems construction and the positional relationship of the cathode, anode and the cooling system. One version suggests using a heat pipe for cooling. It is absolutely new decision for miniature X-ray tubes (figure 4).

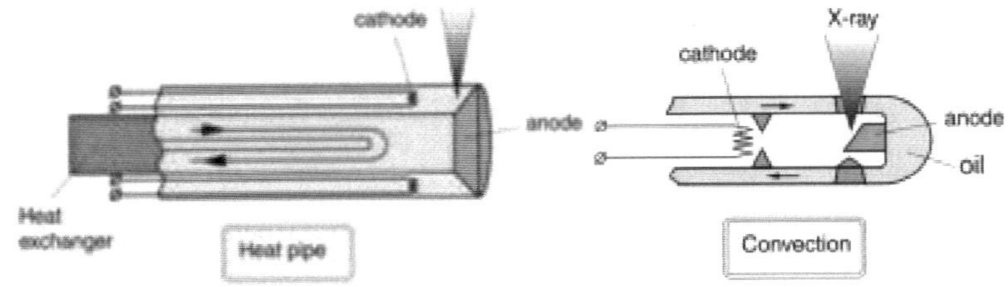

Figure 4. Different cooling systems.

What was the main problem? The first – it needed to have sufficient electron flux density in the confined space. Secondly, anode should have taken increased unit load without damages in spite of its small size. These problems using new materials for cathode and anode that had not been used before for X-ray tubes production had solved. This way cathode and anode can be produced from monocrystal alloy or tungsten nanocomposite. The envelope can be produced of aluminium oxide. All these material are produced at institute of the authors.

Results of development and creation of nanocompozites of refractory metals with increased strength characteristics for anodes and emissive parameters cathodes are analyzed in this report too.

It is shown, that though monocrystals have shown the better characteristics in comparison with traditional polycrystals both as anodes and as cathodes (figures 5–8), more radical way of increase of operational characteristics of the miniature X-ray tubes is in application of the refractory nanostructural materials.

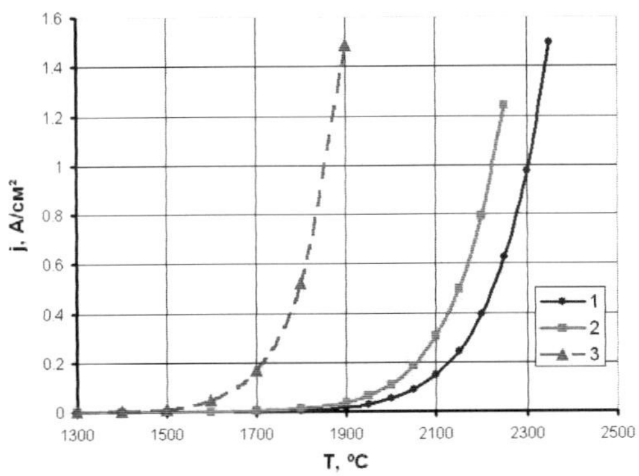

Figure 5. Dependence of emission current: 1 – polycrystalline tungsten; 2 – single-crystal tungsten from plane [111]; 3 – nanocrystalline tungsten.

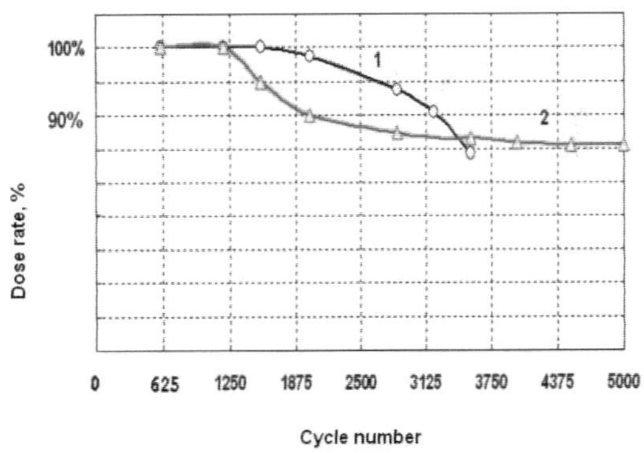

Figure 6. Dose power dependence on the number of switches: 1 – polycrystal tungsten doped by rhenium; 2 – single-crystal tungsten.

Application of nanocrystal and monocrystal materials allows increase allowable unit loads of a miniature X-ray tube by reducing porosity, increasing strength and plasticity of anode and cathode materials. Some results of investigations in which the work function of nanocrystal tungsten is lower by 0.8 eV than that of polycrystals tungsten are given (figure 5, curve 3). This effect is explained by the fact that when forming nanomaterials current tubes with decreased work function arise close to grain boundaries in the region 10 nm wide. This effect is able to increase the intensity of X-radiation considerably (by a factor of more than 5) at the same cathode temperatures or decrease the temperature of a cathode by 400 °C at the same intensity of X-radiation.

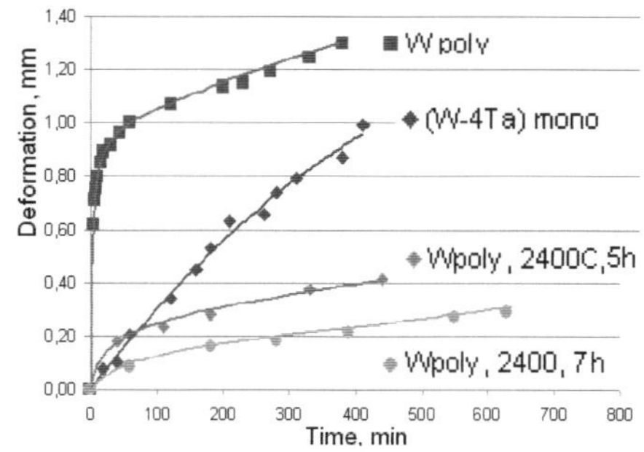

Figure 7. Dependence of deformation on time.

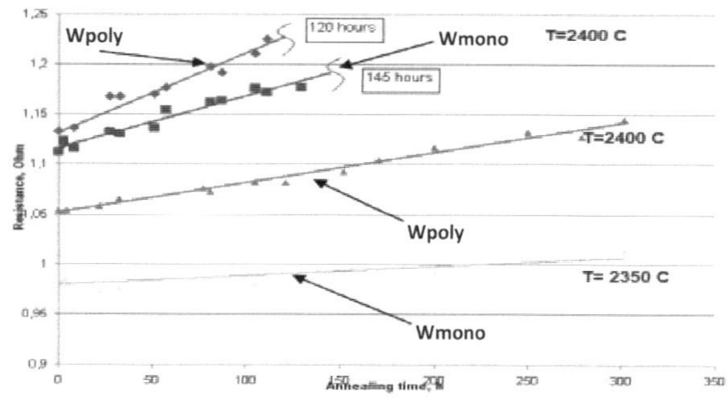

Figure 8. Evaporation of poly- and mono-tungsten.

Authors carried out some experiments to determine the creep of tungsten alloying by potassium on samples as foils 100 μm thick, used when manufacturing X-ray tube cathodes. The results of these experiments showed that high-temperature sample annealing resulted in a increase (by a factor of more than 3) in material creep resistance at a temperature of 2200 °C (figure 7).

Figure 9. Nano scale inclusions on grain.

The authors considers that this effect results from forming nanoscale little bubbles filled with potassium at the grain boundaries (figure 9). Scheme and photo of miniatures of sources for intracavitary thermo radiotherapy with thermal emission are represent (figures 10, 11).

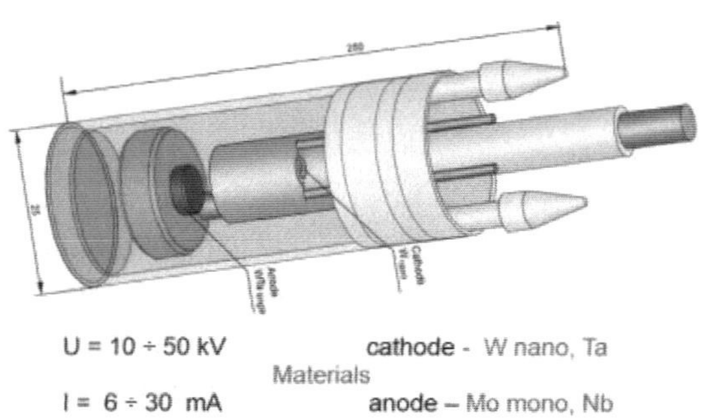

Figure 10. Scheme of miniature source.

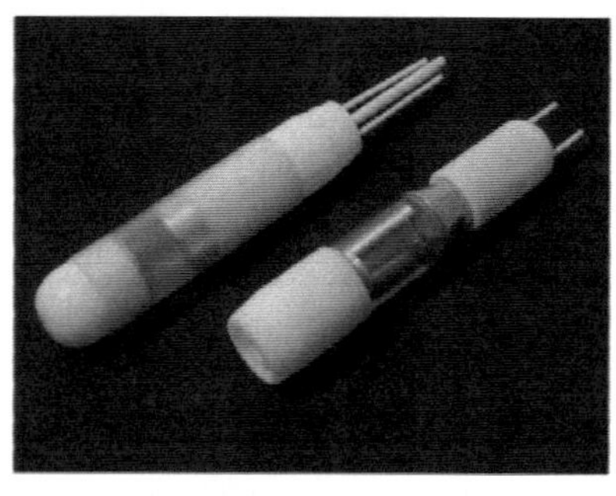

Figure 11. Photo of the miniature sources.

A miniature source of irradiation for thermoradiotherapy produced from monocrystal and nanocrystal materials with a controlled cooling system has been developed. This source can be placed inside the organism nearby or directly at the tumor. This source allows to influence directly the tumor inside the organism by X-ray and heat radiation sequentially or simultaneously.

References

[1] Alekseev S, Taubin M and Yaskolko A 2009 *Proceeding of 20th Workshop in Korea "Nanomaterals and nanotechnology" (Seoul)* pp 46–51

[2] Alekseev S, Taubin M and Yaskolko A 2009 *20th Workshop ISTC in Korea "Nanomaterals and nanotechnology" (Seoul)* pp 52–4

[3] Alekseev S V, Pavlov A A, Taubin M L and Yaskolko A A 2010 *24th ISTC Korea Workshop on "Nano and nano-bio Characterization"* pp 46–9

[4] Alekseev S, Taubin M, Shestikh D and Yaskolko A 2011 *Proceeding of V All-Russian National Congress of Radiologic diagnostics and therapy* pp 20–1

[5] Alekseev S, Ignatiev D, Pavlov A, Taubin M, Shestih D and Yaskolko A 2011 *Collection of articles International conference "Nanostructure in condense medium" (Minsk)* pp 208–14

[6] Alekseev S, Taubin M and Yaskolko A 2012 *Proceeding of the 98th Scientific Assembly and Annual Meeting of the Radiological Society of North America (Chicago)*

[7] Alekseev S, Taubin M and Yaskolko A 2013 *Proceeding of 2nd Biotechnology World Congress (Dubai)*

[8] Taubin M L, Pavlov A A, Chesnokov D A and Yaskolko A A 2017 *IOP Conference Series: Materials Science and Engineering* **175**

IOP Conf. Series: Journal of Physics: Conf. Series **967** (2018) 012018 doi:10.1088/1742-6596/967/1/012018

Existence conditions for a low-pressure high-current discharge in a cylindrical magnetron

S Yu Udovichenko[1], A A Lisenkov[2] and D K Kostrin[2]

[1] Department of experimental physics and nanotechnology, research and education center "Nanotechnology", Tyumen State University, 625003, Tyumen, Russia
[2] Department of electronic instruments and devices, Saint Petersburg Electrotechnical University "LETI", 197376, Saint Petersburg, Russia

E-mail: udotgu@mail.ru

Abstract. In this paper an analytical model of a high-current form of a low-pressure glow discharge in a cylindrical magnetron is presented. Conditions of discharge self-sustainment, allowing to obtain a quantitative estimation of the discharge voltage were found. Were defined the critical values of the magnetic field and residual gas pressure below which the existence of this type of discharge is impossible. It is shown that the emission of fast electrons from the discharge gap changes the conditions of the discharge.

At present time for modification of materials are widely used beams of charged particles generated by plasma of a low-pressure gas discharge [1–3]. To the most promising technological sources of charged particle beams should be attributed plasma sources with magnetron [3–6] and hollow [7] cylindrical cathodes. Due to the oscillations of electrons in a magnetic field or an electrostatic trap discharges in such sources of charged particles exist at low pressures, providing a high degree of ionization of plasma gas and the stability of plasma. Reduction of gas pressure in a glow discharge with cold magnetron and hollow cathodes allows significantly increasing the electric strength of accelerating gap.

A physical model of high-current glow discharge with cylindrical hollow cathode is presented in work [8]. For discharge in the cylindrical magnetron such physical model is missing. The aim of this work is to develop an analytical model of a high-current discharge in cylindrical magnetron, taking into account emission of charged particles from the discharge gap, and to determine minimum values of the magnetic field B and residual gas pressure p at which the existence of this form of discharge is possible.

High-current discharge is modelled in a discharge cell, which is a hollow cylinder of radius R, acting as anode, on the axis of which is located cathode in a form of a rod with radius r_0. A uniform magnetic field B is directed along the axis of the cylinder. In this work is investigated a stationary high-current mode in which the voltage drop is concentrated in a thin near-cathode layer, and the interelectrode space is occupied by quasi-neutral plasma.

A stream of fast electrons generated in the cathode layer, is fed at a speed $V_f(R) = V_{fR} = (2eU/m)^{1/2}$ into the plasma region, where it maintains the necessary degree of ionization. Slow electrons of the plasma not capable to ionize atoms, provide short circuiting of discharge current to the anode.

For the solution is considered a one-dimensional problem in the cylindrical coordinate system. In determining the conditions of the discharge existence, assume that the movement of fast electrons from the cathode to the anode occurs in a diffuse mode, while the influence on the movement of fast particles of a superimposed electric field in quasi-neutral plasma is neglected. Using the stationary equations of continuity and the equations for the flow of fast electrons for radial distribution of particle concentration n_f in the discharge gap [9, 10] we will get the following equation:

$$\frac{1}{\rho}\frac{d}{d\rho}\left(\rho\frac{dn_f}{d\rho}\right)-n_f=-\frac{\gamma n_{ic}V_s\tau_{rf}}{2(R-r_0)}=-n_{f0}, \quad (1)$$

where $\rho = r/\lambda_f$; $\lambda_f = (D_{fr}\tau_{rf})^{1/2}$; $D_{fr}=eUv_{f0}\big/\left[3m_e\left(\omega_{Be}^2+v_0^2\right)\right]$ – diffusion coefficient; v_{f0} – frequency of elastic collisions of electrons with atoms; ω_{Be} – Larmor frequency of rotation; γ – effective coefficient of ion-electron emission; $\tau_{rf} = eU/(v_iW)$ – characteristic time of fast electrons relaxation; v_i – ionization frequency; W – energy spent by electrons on ionization of the gas atoms; n_{ic} and $V_s = (kT_e/m_i)^{1/2}$ – concentration and velocity of ions coming to the cathode from plasma; T_e – electron temperature; m_i – mass of the ion.

The term on the right side of equation (1) describes the appearance of fast electrons in volume due to γ-processes on cathode under the action of ions of the plasma. The lifetime of ions in the collisionless regime is $\tau_i = (R - r_0)/V_s$, and taking into account collisions with the gas atoms – $\tau_i = (R - r_0)^2/D_i$, where $D_i = 2T_i/m_iv_{i0}$ – diffusion coefficient; v_{i0} –frequency of collisions; T_i – temperature of ions. Solution of the equation (1) is found in the form

$$n_f(\rho)=n_{f0}\left[1-AI_0(\rho)-BK_0(\rho)\right], \quad (2)$$

where I_0 and K_0 – modified Bessel functions of zero order of the first and second kind respectively; A and B – constants.

Distribution (2) is decreasing from the cathode to the anode in contrast to the growing solution from the cathode in the case of the hollow cathode effect in the electrostatic trap of a hollow cylindrical cathode [10].

Taking into account the boundary conditions at the anode $n_f(r = R) = 0$ and at the cathode $(n_fv_f)_{r0} = -D_{fr}(dn_f/dr)_{r0}$, we will receive the calculated expression of the concentration distribution of fast electrons in the discharge gap:

$$\frac{n_f(\rho)}{n_{f0}}=1-\frac{P+2(\rho_R-\rho_{r0})G}{P}, \quad (3)$$

where $P = I_0(\rho_R)K_1(\rho_{r0}) + I_1(\rho_{r0})K_0(\rho_R)$; $G = I_0(\rho_R)K_0(\rho_{r0}) - I_0(\rho_R)K_1(\rho)$; $\rho_R = R/\lambda_f$; $\rho_{r0} = r_0/\lambda_f$.

The current density of ions at the cathode, given that n_{f0} is defined in equation (1) in the conditions $en_{ic}V_s = j_{ic}$, could be found using formula (3):

$$j_{ic}=\frac{1}{r_0}\int_{r_0}^{R} ev_in_frdr=j_{ic}\frac{\gamma v_i\tau_f}{2r_0(R-r_0)}\left[\frac{R^2-r_0^2}{2}-\frac{\lambda_fR}{P}M-\frac{2(R-r_0)}{P}(RZ-r_0P)\right], \quad (4)$$

where $M = I_1(\rho_R)K_1(\rho_{r0}) - I_1(\rho_{r0})K_1(\rho_R)$, $Z = I_0(\rho_R)K_1(\rho_R) + I_1(\rho_R)K_0(\rho_R)$ – combinations of Bessel functions. From equation (4) it follows that for self-sustaining of a discharge the following condition should be met:

$$1=u\left(\frac{R+r_0}{4r_0}-\frac{R}{2P\rho_{r0}(R-r_0)}M-\frac{R}{Pr_0}Z+1\right), \quad (5)$$

where $u = U/U_0 = \gamma v_i\tau_f$, $b = B/B_0$ – dimensionless variables; $U_0 = W/(e\gamma)$; $B_0 = 1.5(m_eWv_{f0}/v_i)^{1/2}/$ $/[e\gamma(R - r_0)]$; $v_{f0} \approx 2v_i$; $\rho_R = (2.61b/u)R/(R - r_0)$; $\rho_{r0} = \rho_Rr_0/R$; in the case of magnetized electrons when $\omega_{Be}^2 >> v_{f0}^2$.

The dependence of the discharge sustaining voltage on the magnetic field magnitude, following from (5), is shown in figure 1. For calculations the values of the anode radius R = 0.07 m and the radius of the cathode r_0 = 0.06 m are taken from the system of "inverted magnetron" type [13]. It is seen that the minimum value of the magnetic field for which the existence of a discharge is still possible in cylindrical magnetron is more than in planar magnetron.

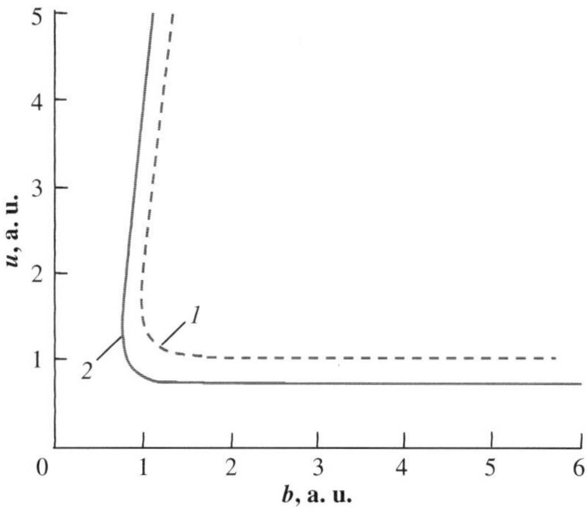

Figure 1. Dependence of the discharge sustaining voltage from the magnetic field induction: *1* – for a planar magnetron with $d = R - r_0 = 0.01$ m [14]; *2* – from equation (5) for the cylindrical magnetron with $r_0/R = 0.86$ ($r_0 = 0.06$ m, $R = 0.07$ m).

The condition of the discharge self-sustaining in a cylindrical magnetron can also be found in another way. As follows from [11], the multiplication factor of the plasma electrons $\alpha = eU/W$, is related with the ratio of the current of fast electrons at the anode to the total current discharge $\Psi = I_{fa}/I$ as follows:

$$\alpha \approx \frac{1}{\gamma - \psi}. \tag{6}$$

Current of electrons to the anode is circuited mainly by the plasma electrons $I = I_{fa} + I_{ea} \approx I_{ea}$, i.e. (6) is valid provided that $\Psi << 1$. Determining the current density of fast electrons from the boundary conditions at the cathode $(n_f v_f)_{r0} = -D_{fr}(dn_f/dr)_{r0}$ and current density of plasma electrons at the anode

$$j_{ea} = \frac{1}{R}\int_{r_0}^{R} e\nu_i n_f r dr, \tag{7}$$

using (6) we will find the condition for discharge self-sustaining in the magnetron:

$$u = 1 + \frac{1}{0.5N\rho_R(1 + 4r_0/R - 5r_0^2/R^2) - 1}, \tag{8}$$

where $N = P[M + 2(\rho_R - \rho_{r0})Z]^{-1}$.

In figure 2 the dependence *2*, following from equation (8), more accurately reflects the conditions of existence of the discharge in comparison with the corresponding dependence *2* (figure 1) because the minimum possible discharge voltage $u = 1$. This condition is achieved when all the fast electrons completely deplete their energy and only after that reach the anode [14].

Presented mathematical model does not allow identifying the effect of residual gas pressure on the voltage of high-current discharge. This can be done, if in equation (6) instead of expression (7) for a current density of plasma electrons at the anode use the expression following expression [11]:

$$j_{ea} = \frac{en_0 V_{Te}}{4\sqrt{1+\beta_e^2}}, \tag{9}$$

where n_0 – plasma density in the interelectrode gap; $V_{Te} = (kT_e/m_e)^{1/2}$ – thermal velocity of the electrons; $\beta_e = 1/(r_{Te} n_g \sigma_{e0})$ – Hall parameter; $r_{Te} = m_e V_{Te}/(eB)$ – cyclotron radius of the plasma electrons rotation; n_g – gas atoms density; σ_{e0} – transport cross section of electron–atom collisions. Then in the assumption of constant concentrations of plasma in the interelectrode gap $n_0 \approx n_{ic}$ instead of (8) we will have:

$$u(1 - \psi/\gamma) = 1, \tag{10}$$

where

$$\frac{\Psi}{\gamma} = 4\beta_e (m_e / m_i)^{1/2} \frac{M/(\rho_R - \rho_{r_0}) + Z}{P}.$$

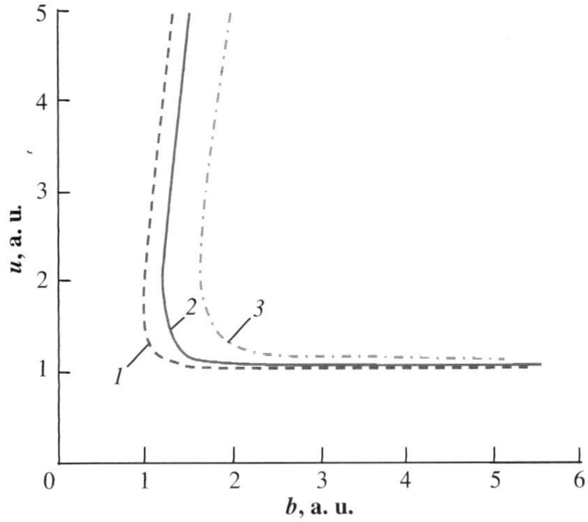

Figure 2. Dependence of the discharge sustaining voltage from the magnetic field induction: *1* – for a planar magnetron with $d = R - r_0 = 0.02$ m [14]; *2* – from equation (9) for the cylindrical magnetron with $r_0/R = 0.7$ ($r_0 = 0.05$ m, $R = 0.07$ m); *3* – from equation (9) for the cylindrical magnetron with $r_0/R = 0.2$ ($r_0 = 0.014$ m, $R = 0.07$ m).

Dependences of the discharge sustaining voltage from the gas pressure (xenon) at various values of the magnetic field induction obtained from equation (10) are shown in figure 3. With a decrease in the magnitude of the magnetic field increases the minimum gas pressure necessary to maintain high-current discharge form.

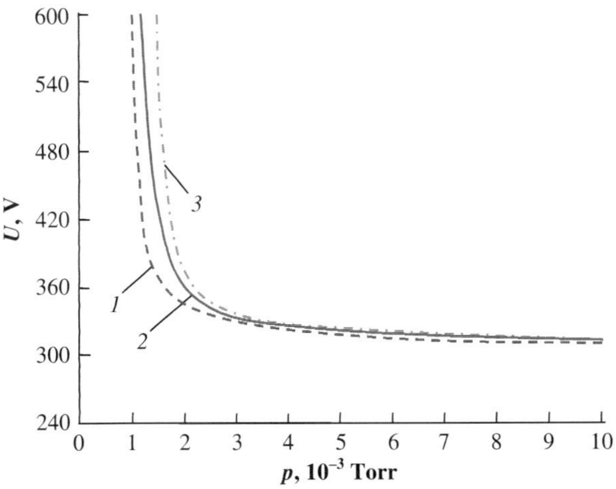

Figure 3. Dependence of the discharge sustaining voltage from the gas pressure (xenon) for different values of the magnetic field: *1* – $b = 1.2$; *2* – $b = 1.0$; *3* – $b = 0.9$. $B_0 = 1.4\cdot10^{-2}$ T, $\gamma = 0.1$, $W = 30$ eV, $\sigma_{e0} = 1.5\cdot10^{-19}$ m^2.

Indeed, the increased departure of the plasma electrons across the magnetic field must be compensated by increasing the intensity of gas ionization by fast electrons. Let us note that in the reflective discharge in cylindrical geometry with a hollow cathode [11] takes place a reverse dependence of the minimum gas pressure on the magnitude of the magnetic field.

The conditions of the discharge are also influenced by the emission of fast electrons from the discharge gap. For the emission of fast electrons from the discharge gap along the magnetic field through the ends of the coaxial cylinder the continuity equation has the form:

$$\frac{1}{r}\frac{d}{dr}\left(rn_f \nu_f\right) = -\frac{n_f}{\tau_{rf}} - \frac{n_f}{\tau_{ef}} + \frac{\gamma n_{ic} V_s}{4(R - r_0)},$$

where the characteristic time of complete loss of fast electrons, according to [14]

$$\tau_f = \frac{\tau_{rf}\tau_{ef}}{\tau_{rf} + \tau_{ef}},$$

where $\tau_{ef} = 2L(2m/eU_c)^{1/2}\varepsilon^{-1}$ – time of particles loss due to the emission from the discharge; L – length of the discharge gap in the direction of the magnetic field (length of cylinder); ε – transparency of the end electrodes in the form of meshes, which are under the anode potential, and through which the emission of particles from the plasma takes place.

In this case, for the discharge self-sustaining the condition (8) must be satisfied, in which the dimensionless parameter u ($u = U/U_0 = \gamma v_i \tau_f$) should be replaced by $u/(1 + \beta u^{3/2})^{1/2}$, where $\beta = \varepsilon(2W/m)^{1/2}/(8\gamma^{3/2} v_i L)$.

The dependence of the voltage of the discharge existence from the magnetic field induction taking into account the emission of fast electrons from the discharge gap is shown in figure 4. With the increase of β the curves are shifted to the right and lower boundary of the magnetic field increases.

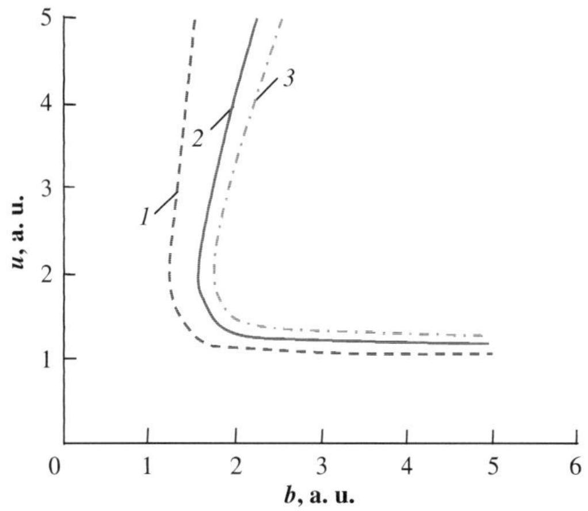

Figure 4. Dependence of the discharge sustaining voltage from the magnetic field induction for $r_0/R = 0.7$: *1* – $\beta = 0$; *2* – $\beta = 0.2$; *3* – $\beta = 0.3$. $B_0 = 1.4 \cdot 10^{-2}$ T, $\gamma = 0.1$, $W = 30$ eV, $\sigma_{e0} = 1.5 \cdot 10^{-19}$ m^2, $\varepsilon = 0.6$, $L = 0.08$ m.

Thus, in the present work, the conditions for maintaining a high-current discharge in the cylindrical magnetron are defined. In this type of magnetron the minimum value of the magnetic field for which the existence of the discharge is yet possible, is higher than in planar magnetron. With increasing distance between the electrodes increases the magnitude of the minimum magnetic field. For the studied geometry the value of the minimum residual gas pressure ($p \approx 10^{-3}$ Torr (0.133 Pa) at $B = 1.7 \cdot 10^{-2}$ T), necessary to maintain high-current form of the discharge is on the order less than the critical pressure for the reflective discharge in cylindrical geometry with a hollow cathode ($p \approx 2.6 \cdot 10^{-2}$ Torr (0.346 Pa) at $B = 2 \cdot 10^{-2}$ T) [11]. With increasing gas pressure decreases the value of the minimum magnetic field needed to maintain a high-current type of discharge in cylindrical magnetron. In the reflective discharge in cylindrical geometry with a hollow cathode this dependency is reverse.

The emission of fast electrons from the discharge gap changes the conditions of the discharge in cylindrical magnetron. With the increase of the parameter β with increasing the transparency of the end electrodes ε and decreasing the length of the cylinder L increases the minimum value of the magnetic field for which the existence of the discharge is possible.

References

[1] Gushenets V I, Hershcovitch A, Kulevoy T V, Oks E M, Savkin K P, Vizir A V and Yushkov G Yu 2010 *Review of Scientific Instruments* **81** 02B303
[2] Semenov A P and Semenova I A 2011 *Proceedings of the higher educational institutions. Physics* **1/3** 217–23
[3] Anders A and Oks E 2009 *Journal of Applied Physics* **106** 023306
[4] Gerasimov V A, Kostrin D K, Selivanov L M, Uhov A A, Komlev A E and Shutova E S 2016 *Journal of Physics: Conference Series* **729** 012019

[5] Kostrin D K, Lisenkov A A and Uhov A A 2016 *Journal of Physics: Conference Series* **735** 012055

[6] Udovichenko S Yu, Kostrin D K and Lisenkov A A 2017 *Journal of Physics: Conference Series* **872** 012013

[7] Boubetra D and Bouafia M 2007 *Journal of American Science* **4** 15–9

[8] Gavrilov N V, Emlin D R, Kamenetskih A S and Menshakov A I 2009 *Proceedings of the higher educational institutions. Physics* **11/2** 85–90

[9] Zhirkov I S, Burdovitsin V A, Oks E M and Osipov I V 2006 *Journal of technical physics* **10** 128–31

[10] Barchenko V T, Krupovich N V and Udovichenko S Yu 2012 *Proceedings of SPbETU "LETI"* **1** 18–23

[11] Kozyrev A V 2009 *Proceedings of the III International Kreyndelevskiy seminar "Plasma emission electronics" (Ulan-Ude)* p 5–9

[12] Zulkova L A, Kozyrev AV and Proskurovskiy D I 2005 *Journal of technical physics* **11** 59–64

[13] Oks E M, Chagin A A and Schanin P M 1991 *Proceedings of the I All-Union conference on plasma emission electronics (Ulan-Ude)* p 18–23

[14] Nikulin S P 1998 *Journal of technical physics* **7** 56–63

IOP Conf. Series: Journal of Physics: Conf. Series **967** (2018) 012019 doi:10.1088/1742-6596/967/1/012019

On the question of use of dosimetric devices in the registration of pulsed X-ray radiation

L V Vladimirov[1], B V Artemiev[1,2], S E Okhrimenko[3] and D M Zashcharinsky[4]

[1] ZAO NIIIN MNPO "SPECTRUM"
[2] Bauman MSTU
[3] RAMPO Department of Radiation Hygiene
[4] Company Canon

Abstract. In this work the features of using the dosimetric devices in the registration of pulsed X-ray radiation are discussed. The dose rate in the pulse mode can reach very high values, while the average value of the dose rate during the operation of the apparatus will be several orders of magnitude lower.

Pulsed X-ray machines are widely used by our industry. This type of apparatus has found wide application in flaw detection, especially when monitoring pipelines in the field, because of their small size and light weight. The duration of the X-ray pulse in accordance with the data of [1] ranges from $2\cdot10^{-9}$ to $1.5\cdot10^{-6}$ s. Safety requirements for pulsed X-ray devices are also regulated by SanPiN 2.6.1.3164-14. It should be noted possible difficulties in the dosimetric control of protection when using pulsed X-ray machines. In the above sanitary regulations, it is indicated that the dosimetric direct-indicating equipment should be designed to measure pulsed radiation. P. 8.9. section "Production Radiation Control" SanPiN 2.6.1.3164-14 [2].

In addition to this point, it is recommended that when dosimetry of pulsed radiation should be used, the integration mode of the dosimeter. In this case, the dosimeter readings in μSv during the measurement period should be divided by the time of inclusion of the X-ray apparatus to obtain the average dose rate during the measurement time. These instructions are correct for the case when the dose rate in the pulse does not exceed the maximum value of the dose rate specified in the dosimeter's passport. It should be remembered that the dose rate in the pulse can reach very high values, while the average value of the dose rate during the operation of the apparatus will be several orders of magnitude lower.

In the case when the dose rate in the X-ray pulse (R_{imp}) exceeds the passport values of the dose rate of the dosimeter ($P_{n.max}$), the error in measuring both the dose rate and the radiation dose becomes uncertain and can reach quite large values. What is most unpleasant, the dosimeter readings at $R_{imp} > P_{n.max}$ can significantly underestimate the actual values of dose rate and dose of X-rays at the site of dosimetric monitoring.

The value of Δt, presented is the time of the X-ray pulse exposure to the detector of the dosimeter device or the time of the X-ray pulse. Accuracy of the dosimeter when measuring the dose rate of pulsed radiation for the case when $R_{imp} > P_{n.max}$ is determined by the following expression:

$$\Delta P = \frac{N\Delta t}{T_{exp}}(R_{imp} - P_{n.max}), \quad (1)$$

where N – number of pulses during exposure time; T_{exp} – exposure time; Δt – duration of one X-ray pulse. The error of the device when measuring the dose during exposure time (T_{exp}) is defined as:

$$\Delta D = N\Delta t(R_{imp} - P_{n.max}), \tag{2}$$

From the analysis of expressions (1) and (2) it follows that the measurement error when switching from the dose rate regime to the dose measurement mode is not eliminated in the case when $R_{imp} > P_{n.max}$. In the dose measurement mode, the measurement error will increase with increasing exposure time or, which is the same thing, with an increase in the number of detected radiation pulses N.

The error in measuring both the dose and the dose rate at the dosimetry of pulsed radiation also depends on the time parameters of the radiation, namely, the ratio of the pulse duration Δt to the pulse repetition period T or the frequency of the pulsed radiation. To confirm this conclusion, we write expression (1) in the following form:

$$\Delta P = \frac{\Delta t}{T}(R_{imp} - P_{n.max}), \tag{3}$$

where $\Delta t = T$; $R_{imp} = P_{const}$, P_{const} – constant dose rate of ionizing radiation; $R_{imp} \leq P_{n.max}$; $\Delta P = 0$ and $\Delta D = 0$.

The dosimeter readings do not give the value of the dose rate in the pulse R_{imp}, but the average value of P_{aver} over a certain time interval T_{tk}, which is determined by the clock frequency of the controller or processor built into the dosimeter.

$$f\,(\text{part of the dosimeter}) = 1/T_{tk}. \tag{4}$$

In this case, the clock speed of the dosimeter does not usually coincide with the repetition rate of the pulses of the X-ray apparatus $T_{tax} = T$, which leads to additional errors in the dosimetry of pulsed radiation by direct-indicating dosimeters. The discrepancy between the clock frequency of the dosimeter and the repetition rate of X-ray pulses is not so critical in estimating measurement errors and is not considered in this article.

In order to be assured of the dosimetric device readings with pulsed radiation, it is necessary to check compliance with the law of reduction or increase of the dialed dose or dose rate in accordance with the change in the square of the distance from the radiation source to the detector of the dosimeter. Make sure that the condition:

$$(\text{readout})\ R_1/(\text{readout})\ R_2 = R_2^2/R_1^2. \tag{5}$$

where R_1 and R_2 – distance from the focus of the pulsed X-ray tube to the point of dosimetry control. The accompanying documentation on X-ray machines with pulsed X-ray tubes often does not mention dose values and dose rate in the pulse. The absence of these parameters is due to the lack of metrological support for measuring the dose rate and dose in the pulse.

However, in order to determine the correctness of the choice of a dosimeter for monitoring the radiation situation, we need to know the dose rate in the pulse and compare it with the technical characteristics of the dosimeter used for dosimetry of pulsed radiation.

An example of calculating the dose rate in a pulse. In the technical documentation for pulse devices, the following parameters are given:

– generation voltage – 100 kV;
– pulse repetition rate (number of pulses N per second) – 40 Hz;
– average value of the anode current is i_{aver} at a frequency of 40 Hz – 0.3 mA;
– radiation pulse time is the same t_{imp} – $2\cdot 10^{-8}$ s.

We determine the dose rate in the pulse at a distance of 1 m from the target of the X-ray tube.

1. The magnitude of the electric current of the X-ray tube during the pulse time:

$$i_{imp} = i_{aver}/N\cdot t_{imp} = 0.3/40\cdot 2 - 8 = 3.75\cdot 105 \text{ mA}. \tag{6}$$

2. From the reference literature [1] we find that the dose rate of constant radiation at a generating voltage of 100 kV and a current of 1 mA at a distance of 1 m from the focus of the X-ray tube, for a 1 mm Al filter is $P = 120$ P·m^2/h·mA.

3. Determine the calculated reduced dose rate in the pulse:

$$P_{imp} = P = 120\times3.75\times105 = 125\ \text{Gy}\cdot\text{m}^2/\text{s}. \quad (7)$$

The technical documentation for dosimetry devices usually indicates a range of dose rates, at which the accuracy of its readings is guaranteed [3]. For example, the well-known dosimeter DRK-1 ensures the correctness of the readings at the maximum dose rate not higher 0.1 Gy/s, which is significantly lower by 4 orders of magnitude than the dose rate in the pulse of the X-ray apparatus at a distance of 1 m from the focus of the X-ray tube. Hence the conclusion: DRK-1 can't be used for dosimetry of the pulsed radiation of the hypothetical apparatus considered in our example.

Consider the possibility of using a personal dosimeter DRK-05D in our case. According to the passport data of the device, the maximum ambient dose rate at which the correctness of its operation is guaranteed is 10 Sv/h. Consequently, this device can't also be used for dosimetric monitoring of a specific pulsed X-ray apparatus.

If, based on the results of the dose rate analysis in the pulse of the X-ray apparatus, we cannot find the required dosimetry device, i.e. $P_{n.max} < R_{imp}$, it is necessary to use thermoluminescent detectors, in which the dose rate limit is not specified. When using a thermoluminescent detector, it should be installed at the point of dosimetric monitoring. Then turn on the device and at a specified frequency determine the exposure time or directly on the device to set the number of radiation pulses. Then, according to the readings taken from the TLD detector, we determine the dose per pulse and, comparing with the allowable dose for the corresponding population groups, we determine how many pulses are permissible for the year of operation of this apparatus. According to the TLD readings of the detector and the permissible annual dose, stationary or mobile radiation protection as well as the boundaries of the sanitary protective dose are also calculated.

For our case, the dose per pulse at a distance of 1 meter from the focus of the X-ray tube is 6.25 μG·m^2/pulse, without the object of transmission. When the aluminum sheet with a thickness of 20 mm is inspected, it takes at least four minutes or 3600 pulses to register onto the X-ray film with amplifying screens.

Let us consider the possibility of using a radiation dosimeter DKS-AT1123. The minimum pulse width for this dosimeter is 10 ns. In our case, the pulse duration is 20 ns. It seems that this is normal. However, the dose rate in the pulse for the impulse device discussed above is 125 G·m^2/s. At that time, according to the passport data of DKS-AT1123, the maximum dose rate in the pulse is 1.3 Sv/s. Let us find the distance from the focus of the X-ray tube of our hypothetical apparatus, in which there will be the possibility of using the dosimeter DKS-AT1123: $R = \sqrt{125} / 1.3 = 9.81$ m, i.e. for this particular case, the distance from the focus of the X-ray tube to the detector of the dosimeter DKS-AT1123 should be at least 10 m.

Thus, for the purposes of dosimetry of pulsed radiation, it is necessary to use it devices after a thorough analysis of its technical characteristics and characteristics of the controlled X-ray apparatus.

An analysis of the technical characteristics of dosimetric monitoring devices for recording the dose and dose rate of pulsed radiation in real time has shown that the most suitable for this purpose are devices that use ionization chambers as radiation detectors. Such a device is the universal dosimeter DKS-101 [3]. Device DKS-101 with ionizing chamber BMK-500 provides registration of photon radiation in the energy range from 0.04…10 MeV with the absorbed dose rate from (0.05…100) Gy/s, which allows recording the dose rate in a direct beam of pulsed X-ray apparatus type "Shmel-90K" produced by the firm Flash Electronics at a focal length of not less than 1.5 m.

When monitoring the radiation situation at workplaces of personnel, as well as the public, a DKS-101 device with the ionization chamber BMK-500, which registers the dose rate in the range (0.006…12) μG/s, should be used.

Analytical expressions for the determination of errors in the measurement of dose rate and dose of pulsed X-ray radiation are presented. It is shown that the use of a direct-indicating dosimeter in the dose-integration mode does not exclude errors in the dose measurement if the dose rate in the pulse exceeds the permissible dose rate, which is indicated in the technical characteristics of the dosimeter.

IOP Conf. Series: Journal of Physics: Conf. Series **967** (2018) 012019 doi:10.1088/1742-6596/967/1/012019

Specific examples of the possibilities of using a number of dosimeters for measuring pulsed radiation are considered.

It is shown that the dosimeter DKS-101 with ionization chambers BMK-50 and BMK-500 can be recommended for monitoring the dose rate and dose of pulsed X-ray radiation, both in a direct beam of radiation, and for protection at workplaces of personnel in real time.

References

[1] Klyuev V V, Muzhitskiy V F, Gorkunov E S, Vladimirov L V, et al. 2010 *Nondestructive testing. X-ray testing*

[2] *SanPiN 2.6.1.3164-14. Hygienic requirements for ensuring radiation safety in X-ray flaw detection*

[3] *Catalog "Radiation monitoring equipment" of the Research and Production Enterprise "Doza"* (Zelenograd)

IOP Conf. Series: Journal of Physics: Conf. Series **967** (2018) 012020 doi:10.1088/1742-6596/967/1/012020

Optimization of the design of scattered radiation detectors

L V Vladimirov[1], I B Artemyev[2] and B V Artemiev[1,2]

[1] ZAO NIIIN MNPO "SPECTRUM"
[2] Bauman MSTU

Abstract. The article deals with the problem of implementing one-way object control with the help of X-ray radiation. Various geometries and materials of radiation detectors are considered. The efficiency of detection of scattered radiation by means of inorganic and organic scintillation detectors with various effective atomic numbers is estimated. Calculations and experiments to determine the signal-to-noise ratio for scanning systems with registration of the radiation beam reflected from the object were carried out taking into account dynamic blurring. Methods for manufacturing detectors that increase the collection efficiency of secondary low-energy quanta formed in the body of the scintillator and reduce the overall dimensions structure are considered. The conditions for achieving the maximum of the efficiency of detection of light photons are determined and the designs proposed for their practical realization.

Certain tasks that have arisen in industry and Russian law enforcement structures have required the creation of X-ray scanners on backscattered radiation – devices that provide information on the internal structure of a controlled object by X-ray radiation when the source and receiver are located on one side of it [1, 2]. When developing this device, recording back scattered radiation from the object of control, it was necessary to solve the following tasks: ensure the required reliability of the registration of the presence of inhomogeneities in the field of control; provide the minimum possible dose load, both to the object of control, and to the operator; provide high performance control.

The reliability of registration is determined both by the deterministic and stochastic parameters of the radiation detection system. Deterministic parameters include: density of the control object and its effective atomic number; density of the inclusion to be detected, its dimensions and the effective atomic number; energy of incident and reflected radiation beams; distance from the focus of the X-ray tube to the target; distance from the monitoring object to the entrance window of the backscattered radiation detector; area of the scanning radiation beam on the surface of the monitoring object; area of the entrance window of the scattered radiation detector; speed of the scanning beam's movement along the object of control; albedo of the scanning beam from the object of the control; albedo of the scanning beam from the inclusion material to be detected.

The stochastic parameters for recording backscattered quanta include: number of quanta falling on the object of control; number of quanta reflected from the object of control; probability of interaction of incident quanta with the material of the object of control; efficiency of registration of reflected radiation by an X-ray detector. The probability of detecting a foreign inclusion in the object is determined by the signal-to-noise ratio.

$$S/N = \Delta N / \sqrt{N_1 + N_2}, \qquad (1)$$

where $\Delta N = |N_1 - N_2|$, N_1 and N_2 – the number of quanta recorded by the detection system from the monitoring object and the artifact (inclusion) to be detected, respectively. When detecting X-ray radiation reflected from an object, the signal-to-noise ratio is determined by the following expression:

$$\left(\frac{S}{N}\right)_{\text{st}} = 2.4\cdot 10^3 \cdot \exp\left[-\left(\mu_{\text{ob}} + \mu'_{\text{ob}}\right)\cdot d\right]\cdot\frac{|\alpha_1 - \alpha_2|}{F_1\cdot F_2}\cdot\sqrt{\frac{P_{\text{o}}\cdot i_{\text{a}}\cdot S_{\text{s}}\cdot S_{\text{d}}\cdot\Delta t\cdot\eta}{\alpha_{\text{av}}\cdot E\cdot\mu_{\text{m}}}}, \quad (2)$$

where μ_{ob} and μ'_{ob} – energy attenuation coefficients in the monitoring object for direct and back scattered radiation, respectively; d – depth of occurrence of the inclusion to be detected; α_1 and α_2 – albedo of the control object material and the inclusion material, respectively; F_1 and F_2 – distance from the focus of the tube to the object of control and from the object of control to the entrance window of the backscattered radiation detector; P_{o} – radiation yield of X-ray radiation; i_{a} – anode current of the X-ray tube; S_{cn} – area of the scanning spot; S_{d} – area of the entrance window of the radiation detector; E – energy of the scanning radiation beam; μ_{m} – mass coefficient of electron conversion of the energy of the scanning radiation beam in air; Δt – scanning time of a single quasi-pixel, determined by the area of the scanning beam and the scanning speed; η – detection efficiency of the backscattered X-ray beam by the detector.

With a signal-to-noise ratio > 3, the probability of recording the foreign inclusion is $P > 99.6$ %. The efficiency of registration when scanning an object of control will be determined not only by the efficiency of absorption of radiation quanta and the efficiency of conversion of quanta into an electrical signal, but also by the speed of the recording system [3].

Expressions (1) and (2) that determine the signal-to-noise ratio do not take into account the dynamics of the registration process when scanning an object of control. In order to take into account the influence of the time for the movement of a narrow X-ray beam on the object of monitoring and the response of the detector signal, we introduce into the expression (2) an additional term determining the dynamic blurriness of the image of the scanning object – K_{d}, and also determine its effect on the signal-to-noise ratio at the output of the registration system.

$$K_{\text{d}} = \exp(\tau/\Delta t), \quad (3)$$

where τ – inertia of the detection system in seconds; Δt – scan time of one quasi-pixel per second.

$$\Delta t = \sqrt{S}/V, \quad (4)$$

where S – area of one quasi-pixel, determined by the aperture of the scanning beam and the distance from the emitter to the scanning object in cm^2; V – velocity of the scanning beam in the plane of the scanning object in cm/s. Taking into account expression (4), the expression for dynamic blur becomes:

$$K_{\text{d}} = \exp\left(V\cdot\tau/\sqrt{S}\right). \quad (5)$$

For the convenience of analyzing expression (5), we expand the exponential function in a series and consider the first two terms of this series:

$$K_{\text{d}} = 1 + V\cdot\tau/\sqrt{S}. \quad (6)$$

The task of maintaining a high signal-to-noise ratio when scanning a monitoring object will be performed at $K_{\text{d}} \to 1$, that is, when $V\cdot\tau/\sqrt{S} \to 0$.

Given that the scanning speed V is set and determined by the performance of the monitoring system, it is desirable to have the maximum possible speed, while maintaining the specified image quality. Increase the area of the scanning beam S is also undesirable, since this leads to a deterioration in the resolving power of the scanner.

There remains one choice – to ensure the maximum speed of the detector, that is, for $\tau \to 0$, $K_{\text{d}} \to 1$. This condition meets the requirement of maintaining a given signal-to-noise ratio at high resolution and a given control performance.

Taking into account the dynamic blurriness, the signal-to-noise ratio for scanning systems with registration of the radiation beam reflected from the object takes the form:

$$\left(\frac{S}{N}\right)_{\text{d}} = \left(\exp\frac{V\cdot\tau}{\sqrt{S}}\right)^{-1}\cdot\left(\frac{S}{N}\right)_{\text{st}}, \quad (7)$$

where $(S/N)_{st}$ – value obtained for the static model by expression (2).

Thus, we were faced with the task of developing a detector with high speed and a high efficiency of detecting X-ray quanta reflected from the object of scanning. When choosing the material of the scintillator conjugated with the photoelectric multiplier (PMT), it was first of all necessary to ensure its high speed. This requirement is best met by organic scintillators: stilbene and anthracene (table 1).

Table 1. The main parameters of organic scintillators.

Scintillator	Density, g/cm^3	Effective atomic number Z_{eff}	Decay time τ, s	Wavelength of the spectrum, nm
Anthracene	1.25	5.8	$25\cdot10^{-9}$	445
Stilben	1.16	5.7	$7\cdot10^{-9}$	410

However, their low density and small effective atomic number do not provide high registration efficiency. While inorganic scintillators (table 2) such as ZnS(Ag) or $CaWO_4$ provide high registration efficiency due to high density and high effective atomic number Z_{eff}.

Table 2. The main parameters of inorganic scintillators.

Scintillator	Density, g/cm^3	Effective atomic number Z_{eff}	Decay time τ, s	Wavelength of the spectrum, nm
ZnS(Ag)	4.1	27	$>10^{-6}$	450
$CaWO_4$	6.06	59	$>10^{-6}$	430
NaJ(Tl)	3.67	50	$2.5\cdot10^{-7}$	410

The widely used inorganic scintillators NaJ(Te) and CsJ(Te) have a decay time of 250 ns and 700 ns, respectively. Which is also significantly more than anthracene and stilbene.

However, for scanning systems, the use of inorganic scintillators leads to image blurring. Therefore, for scanner systems, practically inertial scintillators of anthracene or stilbene type ($\tau \sim 10^{-8}$ s, table 1), should be used. The disadvantage of organic scintillators is their relatively low efficiency of converting the energy of X-ray quanta into light. This is due to their relatively low density, as well as an insignificant effective atomic number (table 1) compared to inorganic scintillators (table 2). In order to increase the efficiency of the conversion of organic scintillators, the technologies for manufacturing scintillator materials with additions of high-atomic metals, for example, tin, have been used [2]. This made it possible to substantially increase the efficiency of recording inorganic scintillators without impairing their speed. Organic scintillators based on polystyrene and polyvinyltoluene with tin additives have a maximum of light output in the ranges 380–425 nm and 370–500 nm, respectively and comparatively high refractive index $n_1 = 1.52$ and $n_2 = 1.58$, respectively, which allows a good collection of light photons at comparatively long scintillation blocks, since they become light guides for formed in them photons of light.

Light photons are formed in the scintillator and propagate in a solid angle of 4π, from which they can either directly enter the photocathode of the photomultiplier, or undergo a complete internal reflection from the faces of the scintillator once or repeatedly. Figure 1 shows a simplified scheme of this process, with the formation of scintillations on the front face and the condition of their total internal reflection, where n – refractive index of the scintillator; α – angle of incidence of the light photon on the opposite face of the scintillator; β – solid angle, within which the formed light photons do not undergo complete internal reflection.

Let us determine the value of the critical angle α, at which the total internal reflection of light photons is achieved: $\alpha = \arcsin(n_2/n_1)$, where n_1 – refractive index of light in the scintillator material; n_2 – refractive index of light in the material surrounding the scintillator. The condition of total internal reflection is achieved for $n_1 > n_2$. For the case when the scintillator is in contact with air, $n_2 \sim 1$ and the condition of total internal reflection is reached when $\alpha \geq \arcsin(1/n_1)$. For scintillators based on polyvinyltoluene $n = 1.58$, $\alpha = 39.3°$. At the same time, the dead angle $\beta = 2\alpha = 78.5°$.

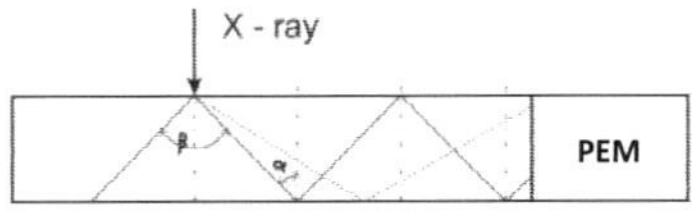

Figure 1. Simplified scheme of the process.

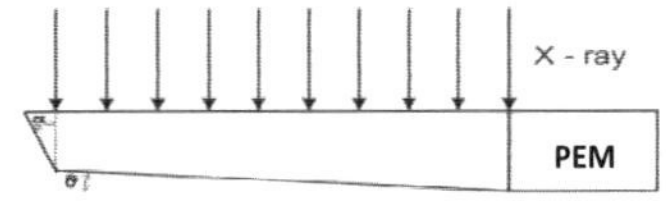

Figure 2. Way to increase the light collection.

In order to increase the light collection of the photon scintillator reflected from the faces, a scintillator configuration with a slope of its faces in the direction of the photomultiplier of the photomultiplier at the angle θ (figure 2) was proposed, which allows to reduce the dead zone β by the values of the slope angle θ and to provide an increase in the light-gathering photon, where θ – angle of inclination of the reflecting face of the scintillator; α – angle of the bevel of the end of the scintillator opposite to the photomultiplier, equal to the critical angle. At $\theta \geq \arcsin(1/n_{sc})$, all photons will be reflected from the opposite face of the scintillator by the condition of total internal reflection. The probability of reflection is given by

$$\eta_o = 1 - 2\arcsin\left(1/n_{sc}\right)/180^o. \quad (8)$$

Expression (8) does not take into account the transparency of the scintillator to its own radiation. The condition $\theta \geq \arcsin(1/n_{sc})$ practically limits the length of the scintillator, contradicts the requirement of the maximum area of its surface S_d, on which the X-ray beam falls (2).

Taking into account the fact that in order to reach the photocathode plane of the photomultiplier, the reflected photons have to undergo multiple reflection from opposite sides of the scintillator, and at each reflection losses take place, taking into account which expression (8) takes the form

$$\eta_1 = C^{K_{otr}} \cdot \left(S_{fk}/S_{ttsc}\right) \cdot \eta_o, \quad (9)$$

where C – reflection coefficient, taking into account the quality of polishing of the faces of the scintillator (~0.9); S_{fk} – photocathode area of the photomultiplier; S_{ttsc} – area of the end face of the scintillator, conjugate with the photomultiplier; K_{otr} – number of reflections before the PEM photocathode hits the plane. Thus, for the critical angle determined by the refractive index of the scintillator and triple reflection with a reflection coefficient of 0.9, the efficiency η_1 is

$$\eta_1 = 0.9^3(1 - \left(2\arcsin\left(1/1.58\right)\right))/180°) \cdot 100\ \% = 41\ \%\ \text{to}\ \left(S_{fk}/S_{ttsc}\right) = 1.$$

At the end position of the photomultiplier to the scintillation unit (figure 3), in addition to the reflected photons reaching the photocathode plane, photons of direct visibility also fall on its surface.

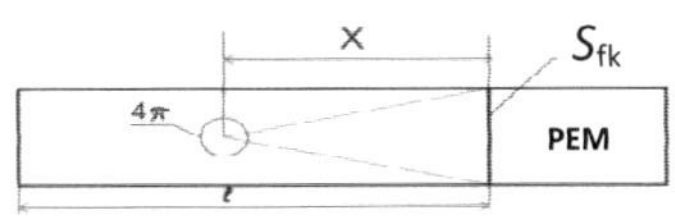

Figure 3. Simplified scheme of the process.

The efficiency of direct entry of photons onto the photocathode of a photomultiplier is determined by the following expression

$$\eta_{nn} = S_{fk}/4\pi X^2 \cdot \exp\left(-X/L_n\right), \quad (10)$$

where S_{fk} – photocathode area of the photomultiplier; X – distance from the photocathode plane to the site of scintillation formation; L_n – relaxation length, which determines the attenuation of the number of scintillation photons by a factor of "e" in the scintillator material.

The total efficiency of collecting photons formed in the scintillator, taking into account both the reflected photons trapped in the photocathode plane and the line-of-sight photons, is described by the following expression

$$\eta_\Sigma = C^{K_{otr}} \cdot \left(S_{fk}/S_{ttsc}\right)\left(1 - 2\arcsin\left(1/n_{sc}\right)/180^o\right) + \left(S_{fk}/4\pi X^2\right). \quad (11)$$

IOP Conf. Series: Journal of Physics: Conf. Series **967** (2018) 012020 doi:10.1088/1742-6596/967/1/012020

The expression (11) does not take into account the self-absorption of photons in the scintillator material, determined by the path length of the photons to the PEM photocathode and the relaxation length L_n (10). Expression (11) takes into account only the entry of photons onto the entrance surface of the photocathode, without taking into account the absorption and reflection of photons from the input window of the photomultiplier and without taking into account the coefficient of spectral matching of the scintillator photon spectrum and the spectral distribution of the photocathode sensitivity. Let us determine the number of light photons formed in the scintillator in recording X-ray quanta (phot/s)

$$N_{ph} = \left(10^3 \cdot N_{ph} \cdot E \cdot S \cdot \left[1 - \exp\left(-\mu_t \cdot d\right)\right] / W\right) \cdot \left(\mu_e / \mu_t\right), \tag{12}$$

where N_{ph} – number of X-ray quanta incident on the scintillation detector (kv/cm^2·s); E – quantum energy (keV). The energy of the backscattered quantum at U_a = 100 kV corresponds to 30 keV; S – area of the input surface of the scintillation detector (cm^2); μ_t – coefficient of linear attenuation of the scintillator for energy E (cm^{-1}); μ_e – coefficient of electron energy conversion in the scintillator material, (cm^{-1}); d – thickness of the scintillator in the direction of the X-ray flux (cm); W – energy expended by the electron formed in the scintillator material to form a single photon (eV). For organic scintillators, $W \sim 32$ eV.

When scanning a control object with a narrow beam of radiation with the beam area at the object input ΔS (cm^2) and the scan rate V (cm/s), the time of irradiation of the scintillator with quanta reflected from one quasi-pixel Δt is determined by the following expression (s): $\Delta t = \sqrt{\Delta S} / V$, Δt can also be set by the clock frequency of the pulse counter at the PMT output. In this case, the number of photons produced in the scintillator during the time Δt is defined as:

$$N_{ph(\Delta S)} = N_{ph} \cdot \Delta t, \tag{13}$$

where N_{ph} is determined from expression (11). The next task is the transmission of light photons to the photocathode photomultiplier.

$$N_{fk} = \left(N_{ph} \cdot S_{fk} / 4\pi\right) \cdot \exp\left(-l / L_\Sigma\right), \qquad (14),$$

where S_{fk} – photocathode area of the photomultiplier; l – length of the scintillator; L_Σ – relaxation length of photons in the scintillator, at which the attenuation of photons occurs "e" times. The length of relaxation depends not only on the scattering and absorption of scintillation photons in the scintillator material, but also on the geometric dimensions, the quality of the polishing of the scintillator faces and the reflection coefficient from the scintillator faces and their number. Expression (14) refers to the case when one photomultiplier is used, located at the end of the scintillator.

Expression (14) determines the number of light photons formed in the scintillator and trapped on the photocathode when the photomultiplier is placed at one end of the scintillation block.

The total number of light photons reaching the photocathode photomultiplier is determined by the following expression

$$\Sigma N_{fk} = \left(N_{ph} \cdot S_{fk} / 4\pi\right) \int_0^l \exp\left(-x / L\right) dx, \tag{15}$$

where l – length of the scintillator. Expression (15) corresponds to the requirements of an ideal light guide with complete internal reflection from the faces of the scintillator and photons incident on the photocathode.

For practical calculations, by expression (15), we can replace the integral from "0" to "l" by a sum of discrete values with a given step along the length of the scintillator. Moreover, the expression (15) without taking into account the constant term $(N_{ph} \cdot S_{fk})/4\pi$ takes the form

$$\sum_1^n N_{fk} = \left(N_1 + N_2 + \ldots + N_n\right), \tag{16}$$

where $N_1 = N_{ph} \cdot \exp\left(-\Delta l / L\right)$, $N_2 = N_{ph} \cdot \exp\left(-2\Delta l / L\right)$, …, $N_n = N_{ph} \cdot \exp\left(-n\Delta l / L\right)$, $n = l/\Delta l$.

The collection of light photons by conjugation of a scintillation crystal with a photomultiplier is optimal under the condition that $S_{fk} = S_{ttsc}$ when the photocathode area of the photomultiplier is equal to the cross-sectional area of the scintillator.

As an example, figure 4 shows a sensitivity diagram of a scintillator with a width of 200 mm and a length of 650 mm, depending on the location of the gamma source along its length.

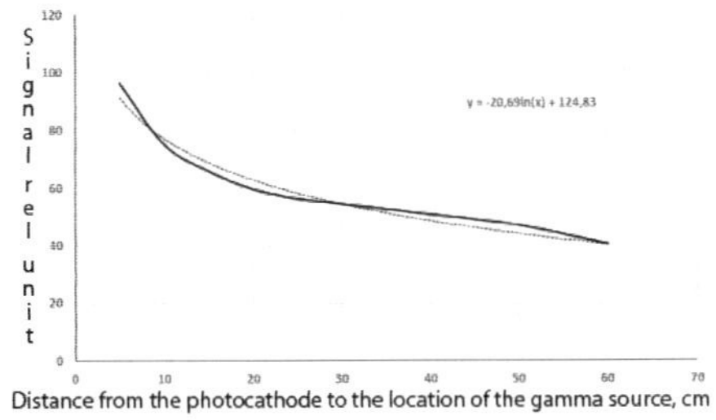

Figure 4. The sensitivity of the scintillator, depending on the location of the gamma source along its length.

For the purpose of registering radiation with a scanning X-ray beam, in order to reduce the dynamic blur, it is necessary to fulfill the condition: $\tau << \sqrt{S}/V$.

The efficiency of registration reaches its highest value, provided that: (photocathode area)/(area of the cross section of the scintillator) → 1. In this case, if possible, a photomultiplier with a maximum photocathode diameter should be used.

In order to increase the collection of light photons, it is possible to skew the opposite end face of the scintillator to a critical angle α, and also to use the bevel of the scintillator at an angle θ in the photocathode photocathode direction.

To smooth out the response function of signals from the photomultiplier when scanning with a narrow scintillator beam, three or more photomultipliers should be used to select photons from the dead band formed by the angle β. The results of measurements are presented in table 3.

Table 3. The results of measurements.

Area of the input surface, cm^2	Thickness, cm	Length, cm	Width, cm	Sensitivity, Nm/cm^2	Efficiency η, %
1270	3.5	63.5	20	0.3	5
825.5	3.5	62	14	0.36	6
23.4	3.5	27.5	8.5	0.6	10
42.3	3.5	27.5	16	0.58	9.6

Also, the dose rate of backscattered X-ray radiation in the plane of the scintillation block input window was measured with the help of the DKS-AT1123 dosimeter. The dose rate at $U_a = 100$ kV and $i_a = 3$ mA was 45 µSv/h. At a scattered photon energy of ~30 keV, the measured dose rate corresponds to 6 ph/cm^2 during the time $\Delta t = 64$ µs.

References

[1] Blokhin E O, Parshin I A, Shurushkin A V, Buckley A A, Artemiev B V and Fedorovsky E V 2012 *Control. Diagnostics* **12** 4–7
[2] Artemiev B V, Bukley A A and Shubochkin A E 2012 *18th World Conference on Non-Destructive Testing*
[3] Bukley A A, Vladimirov L V and Kozlov A A 2009 *Medical technology* **5** 27–30